# Undo the Math!

**Collection « Recherches en Gestion »**
Dirigée par Luc Marco

*Apprentissages, stratégies et compétitivité sur la longue durée. L'étonnante histoire d'Eurocopter*, Marc-Daniel Seiffert, Préface de Jean-Claude Tarondeau, 2008.

*Les courants actuels de recherche en marketing. Synthèses et perspectives*, sous la direction de J.-M. Décaudin, J.-F. Lemoine et J.-F. Trinquecoste, 2006.

*Les coûts des maisons de retraite*, Hien et Miyako Bui Quang, 2008.

*Design et marketing : fondements et méthodes*, sous la dir. de J.-P. Mathieu, 2007.

*L'entrepreneur français, modèle pour le XXI^e^ siècle*, Karine Goglio-Primard, 2007.

*La fabrique des experts-comptables, une histoire de l'INTEC*, Luc Marco, Samuel Sponem et Béatrice Touchelay, 2011.

*Histoire managériale du Bazar Bonne-Nouvelle, galeries marchandes à Paris, 1835-1863*, Luc Marco, 2009.

*Inefficience et dynamique des marchés financiers*, F. Jawadi, J.-M. Sahut, 2009.

*Le management de la diversité. Enjeux, fondements et pratiques*, sous la direction d'Isabelle Barth et de Christophe Falcoz, 2007.

*Méthodes et thématiques pour la gestion des risques*, sous la direction de Bernard Guillon, 3 volumes, 2007, 2008, 2011.

*Morale industrielle et calcul économique dans le premier XIX^e^ siècle. L'économie industrielle de Claude-Lucien Bergery, 1787-1863*, François Vatin, 2007.

*Nouvelles avancées du management*, sous la direction de Luc Marco, 2005.

*La presse et les périodiques techniques en Europe, 1750-1950*, sous la direction de Patrice Bret, Kostas Chatzis et Liliane Pérez, 2009.

*La publicité paneuropéenne*, sous la direction de Sylviane Toporkoff, 2006.

*Regards actuels sur la société contemporaine. La pensée de Georg Simmel*, sous la direction d'Isabelle Barth, 2009.

*Traité de la marchandise et du parfait marchand*, Benedetto Cotrugli, 1582, édition scientifique par Luc Marco et Robert Noumen, 2008.

Marc Idelson

# Undo the Math!

## Managerial and organizational cognition theoretical and practical implications of cross-cultural and cross-disciplinary semiotic gaps

L'Harmattan

Cover photograph : © ESSEC, 2010.
Cover illustration : M. Idelson, 1986, 2011.


**5-7, rue de l'École-Polytechnique ; 75005 Paris**

http://www.librairieharmattan.com
diffusion.harmattan@wanadoo.fr
harmattan1@wanadoo.fr

ISBN : 978-2-296-96993-3

EAN : 9782296969933

*To Professors Bitbol, Crosby, Donaldson,*
*Jorion & Jullien on whose shoulders I stand,*

*To that great device, the anonymous reviewer,*
*for raising my spirits in the nick of time,*

*To Professor Besseyre des Horts*
*for his incommensurable patience,*

*To the memories of Professors Cabrol & Dumont,*
*who first nudged me onto this journey.*

*To my close friends and family,*
*for never asking but always questioning,*

*To fate, for auspiciously putting me within*
*the purview of my then future wife.*

*"When I choose a word," Humpty Dumpty said in a rather scornful tone,*
*"it means just what I choose it to mean – neither more nor less."*
"Through the Looking Glass", Lewis Carroll

# MATHEMATICS STANDS AS THE FIERCEST ENEMY OF SOCIAL CONSTRUCTIONIST ONTOLOGIES,

## BY ERIC ABRAHAMSON[1]

This book turns on a fascinating premise, that Society constructs Mathematics. In other words, Mathematics does not inscribe itself into social consciousness. Mathematics, like any other Reality, is socially constructed.

What type of social construction? Take a man who painfully bumps into a wall. Remember Society calls the man a man and the wall a wall. A Society of different men who call themselves "carpenters" builds the walls Society mandates. Society tells the man how to react when he bumps against the socially constructed Reality he is walled into. Society determines the level and nature of what the man calls "pain" when he bumps into the boundaries Society constructed. Society even creates unbumpable pain barriers that are never tested. In this way, Society constructs the one and only Reality. No one thinks outside the Societally constructed Reality of something that is not socially constructed. Societally constructed Reality has no outside. Nothing stands behind it to influence it.

Mathematics, however, stands at the most obdurate wall that Society arguably does not construct. Mathematics dictates the wall's perimeter, area, and volume —this is beyond social discussion. More pointedly, a circle with a radius r has its own, real properties independent of what Society does or does not construct. It has an area A such that:

$$A = \pi r^2$$

It has a perimeter C, such that:

$$C = 2\pi r$$

---

[1] Professor Eric Abrahamson is the Hughie E. Mills Professor of Business Management at Columbia Business School.

π is a symbol that in base 10 equals 3.14 followed by an infinite series of numbers; π stands outside socially constructed Reality. Moreover, π inscribes itself into Society. It operates independently on whatever round artifact Society decides to construct.

To pass the time at a cocktail party, I found nothing better to do than to propose to a Nobel Laureate in Chemistry that π existed only insofar as it was produced by Society. If a Society constructed nothing circular, I proposed, then there would be no circle areas or perimeters. There would exist no real π to measure these areas and perimeters. More simply, I stated what Mathematicians have stated more eloquently. There is no number 2 without the socially mediated perception that there are two like members of a set. "2-ness" does not exist alone just as "π-ness" does not exist without circles.

The Nobel Laureate looked at me, bemused. I had not even planted in his mind the idea that it might be interesting to consider that π is just as real as any other social construction, like a wall. For him Societies without a π and, by extension, Societies without Mathematics, were Societies that were too primitive to have discovered the Mathematics of π. The recognition of π constructed modern Society. Modern Society did not construct π. π existed before Society, made it possible for Society to develop wheels and for Mathematics to articulate Heisenberg's uncertainty principle. π would outlive Society's self-destruction.

Husserl once wrote that in a completely blue world there would be no blue. Blue becomes real only in contrast to another color. When social construction creates a hegemonic Reality then the hegemon disappears from sight. Likewise the social construction of a western paradigm based on Math so surrounds and suffuses us that we can no longer see that we are surrounded by it. Numbers and Math are no longer what Society constructed to measure Reality; they are that Reality which lets itself be accurately measured by Society; and not just any Society, only an "advanced" one.

This book could create an instant of doubt in the Nobel Laureate's mind.

# Thesis abstract

In a critical management study, drawing on Aristotelian analysis of language[2], Crosby[3]'s historical analysis of the emergence of what he coins "modern society's pentametric *mentalité*", Jorion[4]'s anthropological exposition of the West's invention of "*Truth*" and "*(objective) Reality*", and following, albeit with a much more limited scope, in the footsteps of Hellenist philosopher Jullien[5]'s forays into Chinese thought, I prove (a) numbers, space and time are social constructs and (b) philological methods —revealing semiotic gaps[6]— shed light unmatched by past multicultural surveys[7].

From this, I plan to demonstrate the merits of:

(a) exercizing self-doubt, open-mindedness, and unlearning capability, whilst in cross-cultural and cross-disciplinary contexts,

and (b) awareness of cross-cultural and cross-disciplinary cognitive gaps at the individual, team, group and organizational levels.

---

2 Aristotle. 2000a. *Catégories de l'interprétation: Organon I et II*: Vrin, Benveniste, E. 1966. *Problèmes de linguistique générale*. Paris: Gallimard.

3 Crosby, A. W. 1997. *The Measure of Reality: Quantification and Western Society, 1250-1600*: Cambridge University Press.

4 Jorion, P. 2009. *Comment la vérité et la réalité furent inventées*. Paris: Gallimard.

5 Jullien, F. 1995. *Le détour et l'accès: stratégies du sens en Chine, en Grèce*. Paris: Grasset, Jullien, F. 2009. *L'invention de l'idéal et le destin de l'Europe*. Paris: Seuil.

6 Cruse, A. D. 2004. *Meaning in Language*. Oxford: Oxford University Press, Lyons, J. 1995. *Linguistic Semantics: an introduction*. Cambridge: Cambridge University Press.

7 Hofstede, G. 1983. National Cultures in Four Dimensions: A Research-Based Theory of Cultural Differences Among Nations. *International Studies of Management and Organization*, 13(1,2): 46, House, R. J., Hanges, P. J., Javidan, M., Dorfman, P., & Gupta, V. 2006. *Culture, leadership, and organizations: the GLOBE study of 62 societies*. Thousand Oaks, CA: Sage.

Anchored epistemologically in Popperian falsifiability[8], Kuhnian scientific progress[9] and Porter's sociological treatise on our trust in numbers[10], I join the debate on the nature of social science.

(i) I then explore what my empirical findings reveal in the context of mathematically supported model-building in the realm of social science; acknowledging mathematics as a social construction[11] is, I observe, often ignored in practice by both social scientists and business practitioners;

I evoke how post-Newtonian theoretical physics builds new theory by building new math and describe mathematically sound alternatives to Peano arithmetic and Euclidean geometry that have not been explicitly dismissed or considered by social scientists;

I reconcile independent and superficially paradoxical attempts to theorize organizations with various bodies of mathematics (notably Bitbol's and Donaldson's recent attempts[12]);

and I propose development principles towards a phenomenologically grounded mathematics corpus.

---

8 Popper, K. 2002. *The Logic of Scientific Discovery* (15 ed.): Routledge.

9 Kuhn, T. S. 1996. *The Structure of Scientific Revolutions* (3 ed.): University of Chicago Press.

10 Porter, T. M. 1995. *Trust in Numbers*. Princeton: Princeton University Press.

11 Berger, P. L., & Luckmann, T. 1966b. *The Social Construction of Reality: A Treatise in the Sociology of Knowledge*: Anchor Books, Jorion, P. 2009. *Comment la vérité et la réalité furent inventées*. Paris: Gallimard.

12 Bitbol, M. (Ed.). 2009. *Théorie quantique et sciences humaines*. Paris: CNRS, Donaldson, L. 2010. *The Meta-Analytic Organization, introducing Statistico-Organizational Theory*. Armonk, NY: M.E Sharpe.

(ii) next, I expose how a social network perspective[13] applied conjointly at the semantic and cultural levels of analysis may usefully extend and bring insights to standard qualitative content analysis[14].

To illustrate the practical implications of my thesis, I focus on one fertile ground and propose a loss-less method to merge information systems. Empirically illustrating the Aristotelian breakdown of language into its component parts (and the special role numbers, space, and time play in Western discourse and language), I propose a methodology for non-loss post-merger integration of organizational knowledge that builds both:

(a) on Boyce, Codd et al's database normalization[15] from the field of Information Management in Computer Science;

and (b) on Nonaka's SECI dynamic model of knowledge creation[16] from the field of Knowledge Management in Business Administration Studies.

---

13 Wasserman, S., & Faust, K. 1994. *Social Network Analysis: Methods and Application*: Cambridge University Press.

14 Demazière, D., & Dubar, C. 1997. *Analyser les entretiens biographiques. L'exemple des récits d'insertion.* . Paris: Nathan.

15 Codd, E. F. 1970. A Relational Model of Data for Large Shared Data Banks. *Communications of the ACM*, 13(6): 377–387, Codd, E. F. 1972. Further Normalization of the Data Base Relational Model. In R. J. Rustin (Ed.), *Data Base Systems: Courant Computer Science Symposia Series 6*: Prentice-Hall, Codd, E. F. 1974. Recent Investigations into Relational Data Base Systems (IBM Research Report RJ1385), *Proc. 1974 Congress.* New York: North-Holland, Date, C. J., Darwen, H., & Lorentzos, N. A. 2003. *Temporal Data and the Relational Model: A Detailed Investigation into the Application of Interval and Relation Theory to the Problem of Temporal Database Management.* Oxford: Elsevier, Fagin, R. 1977. Multivalued Dependencies and a New Normal Form for Relational Databases. *ACM Transactions on Database Systems*, 2(1): 267, Fagin, R. 1979. Normal Forms and Relational Database Operators, *International Conference on Management of Data.* Boston: ACM SIGMOD, Fagin, R. 1981. A Normal Form for Relational Databases That Is Based on Domains and Keys. *Communications of the ACM*, 6(3): 387-415.

16 Nonaka, I. 1994. A dynamic theory of organizational knowledge creation. *Organization*

This methodology, coined as archnormalization, introduces the sociologically grounded concept of Attribute-free Entities.

Finally, further avenues of research are proposed, notably that the observed implicit semiotic paucity of the body of theory of organizational learning —in support of which March's interplay of exploration and exploitation are simulated by computer[17]— reveals existing models ignore the heterogeneity of individual cognition structures supporting the learning process —notably including vocabularies, grammars and meaning structures. Remedies —to be jointly developed with Mangin[18]— are hinted at, but ultimately lie beyond the scope of this thesis.

---

*Science*, 5(1): 14.

[17] Fang, C., Lee, J., & Schilling, M. A. 2010. Balancing Exploration and Exploitation Through Structural Design: The Isolation of Subgroups and Organizational Learning. *Organization Science*, 21(3): 625-642, Lazer, D., & Friedman, A. 2007. The Network Structure of Exploration and Exploitation. *Administrative Science Quarterly*, 52(4): 667-693, Levinthal, D. A. 1997. Adaptation on rugged landscapes. *Management Science*, 43(7): 934-950, March, J. G. 1991. Exploration and Exploitation in Organizational Learning. *Organization Science*, 2(1): 71-87.

[18] Idelson, M., & Mangin, N. 2010. Learning in social networks: simulating the impact of different cognitive frames and interpretation schemes on exploration and exploitation. In S. A. Rodan (Ed.), *2010 Academy of Management Simulation in Organization Theory Professional Development Workshop*. Montréal.

# TABLES

## Table of contents

## Tables

## Cases



## Figures

# Foreword

In my first year of middle school, I was fortunate to be taken to our school's computer lab by both my French Letters and Mathematics teachers (this was 1974 and computers in secondary education were fairly rare).

Professor Cabrol would tell our class how wonderful computers were as emerging tools, allowing us to state rules —such as those of Latin conjugations— and have programs apply them in unerring, mechanical perfection.

Professor Dumont would tell our class how wonderful computers were as emerging vessels for our budding, creative minds, enabling us, by programming what we can imagine, to explore and devise pre-existent and new concepts.

The reflexive exercise my thesis has instigated me to start on others' and my own thought processes —and their interplay— owes, I realize now, much to each, but most to being confronted early to both perspectives. The unintentional, concurrent exposure to their viewpoints is, I believe, not unrelated to my worldview-building to this day.

Two years later, our computer lab acquired a graphic terminal and I computed my first animated sequence that same year.

Visually illustrating and animating my conceptual musings by computer hasn't left me since, even if I now have learnt to be as wary of drawing the wrong conclusions from graphical representations as from single-perspective expositions of accounts.

That I situate myself today in the structuralist scholarly stream is, I see today, indomitably tied to these timid first steps in computing, *climbing step by step in the snow up a mountain now accessible by road, rail and cable car*[19].

[19] The credit for this potent image goes to van Dam, A. 2008. Reception address. In Computer Graphics Pioneers (Ed.), *SIGGRAPH 2008*. Los Angeles, CA: Association for Computing Machinery.

Should I state "*I see all as structures*", "*see*", "*all*" and "*structures*" warrant qualifications:

- "*to see*" to me here means as much "*to think*", "*to project*", as it does "*to aesthetically appreciate the form*";
- "*all*", "*anything I perceive*" as well as "*anything I conceive*";
- and "*structures*", that I reflexively tend to link anything newly perceived or conceived within things previously perceived or conceived.

Were I to express all this, formally, I would mobilize set theory and specifically term these observations with elements and relationships, i.e.:

- any newly perceived or conceived thing is an element I instinctively (and perhaps obsessively) attempt to link relationally with previously perceived or conceived elements;
- some of these elements will be members of others elements (which are sets) but more generally elements will be in qualified relationships with others, membership being just one of many possible relationships;
- it is this mesh of multilayered relationships that I mean by structures.

Thus, as a structuralist thinker and a scholar keenly interested in human collective behavior and in mathematics, I have turned my attention to organizations as vessels of imperfectly shared knowledge. It was unavoidable that, sooner or later, I would ponder the topic of mathematics and the behavioral aspects surrounding its application within organizations.

As mathematics objects are purely symbolic, the study of cognitive representations of these objects has led me to draw from the field of semiotics in this PhD in Business Administration thesis.

**Traduttore, traditore**[20]

To expose how important cognitive representations are to collective action, I will turn my focus to that most universal of mathematical constructs, the number, and show, within Sections I-IV, how structural perception of numbers is conditioned by language.

> In section I, the rationale that initiated this line of research will be delineated.
>
> In section II, an epistemological approach and organization theories will be presented and information, communication, language, knowledge and reflexivity discussed, drawing upon the realms of philosophy, logic, sociology, psychology, history, anthropology and economics.
>
> In section III, a proposed empirical approach intending to test the propositions put forward in section II will be outlined.
>
> In section IV, attempts to refute the propositions will be reported upon.

Within this exposition, I also pore over the concepts of space and of time, as the studies of Western languages draws common conclusions on numbers, space and time. How this specificity broadens our understanding of collective knowledge building, representing and sharing has wide-ranging, though not necessarily novel, implications for business research and practice.

From the narrow focus of numbers, I then extrapolate, within Sections V-VI, to all forms of modeling social phenomena from quantitatively-driven theoretical modeling to mere business modeling when designing a new venture. I suggest a few, possibly non-exhaustive preventive and corrective measures to currently flawed social phenomena modeling practices and thus propose a research program.

---

[20] "Translator, traitor." (Italian paronomasia).

In section V, the broader, potential theoretical implications for quantitative modeling will be highlighted.

In section VI, a research program will be proposed to pursue the theoretical inquiry raised in section V.

Drawing from the social nature of information and the institutional and methodical explicitation of the context of its creation, I propose incremental, but potentially rewarding improvements to several academic and business practices.

Section VII suggests a narrower theoretical implication aimed at enriching current qualitative research methodological practices by adding semiotic representation to current methods.

In section VIII, broader, potential managerial implications will be highlighted and the methodical development of practical management tools drawing upon these will be suggested.

In section IX, an approach to information systems mergers is developed as an illustration of an innovative managerial practice.

Finally, section X discusses future research and various ancillary sections follow.

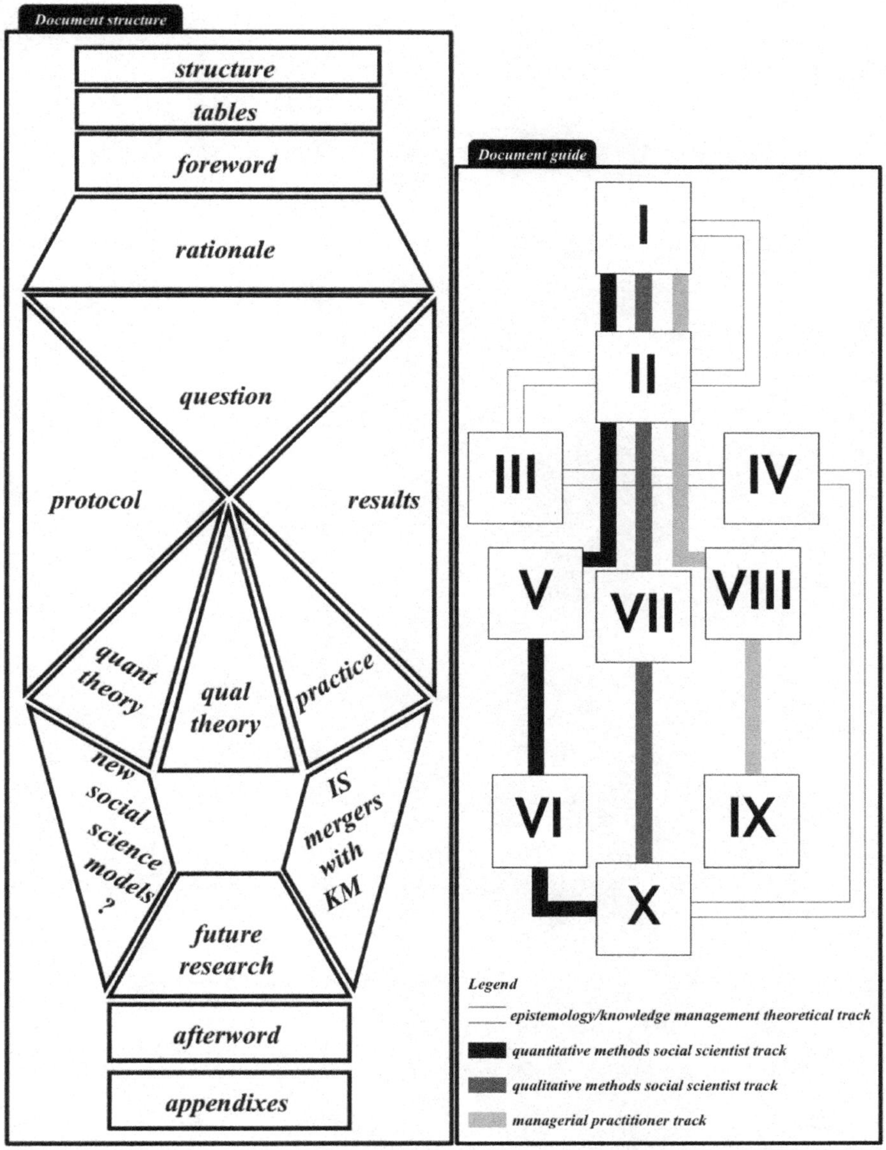


*Figure 1: Document structure and guide*

Readers who do not wish to read the whole document can use the guide in Figure 1 to select which sections to delve into.

# I] RATIONALE —HARD SCIENCE HARDLY HELPING

**i the case of the inverted envelope**[21]

A Chinese student had recently moved to France. On one occasion, he glanced at the envelope a French student was writing and exploded into laughter. "Your envelope is written backwards!"

Chinese envelopes are written from the geographically general to the geographically specific (i.e. Country or Province first, then Town, Street, Building, Floor and Recipient last) whereas Western envelopes first state the Recipient and end with the Country.

Once again, the Chinese write addresses on envelopes with an awareness the envelope is for the postman and seek a frictionless insertion of their letter to the postal flow; on the other hand,

Westerners write the name of the recipient first because that's what's most important to them.

Envelope address writing is means-focused in China, ends-focused in the West.

**ii the case of the disagreeing CxOs**[22]

During the executive committee meeting of the Asia Pacific regional HQ of a German insurer, Pat, the British-born Regional CTO, non-Life presents the latest claim rates analysis. Also present are the Regional CEO (and chair of the committee), COO, CTO, Life, CFO and CIO.

At one stage of Pat's presentation, Sam, the India-born CFO, a former accountant with a reputation for a sharp mind, rigor and accuracy, bluntly interjects: "Your figures are false."

---

21 François Cheng narrated this to the author.

22 This incident was directly experienced by the author; particulars have been anonymised.

A seasoned actuary[23], proud of his ability to crunch numbers, who joined the insurance industry at age 17 and rose through the ranks, Pat doesn't take this interruption kindly —few things can kill a business presentation faster than an accountant stating the incorrectness of presented figures.
Tempers flare with Pat and Sam both staking their professional prides on their respective positions. Claude, the French CEO, sighs, slips to Dominique, the French CIO: « *Dépatouille-moi tout ça !* »[24] then gets up, thereby signalling an early conclusion to the meeting.
Dominique investigates. The specific point of contention is Gross Written Premium in Auto Insurance by the Group in Hong Kong for Q3 of the previous year. After having heard both parties separately, Dominique reports back to Claude: (a) both agree on the definition of Gross Written Premium; (b) both agree on what Auto Insurance is; (c) both agree on the perimeter of the Group in Hong Kong; (d) and the figure was expressed in the local currency so no exchange rate discrepancy is at the origin of Alex and Lee's disagreement; (e) what they disagreed on was … what 'Q3 of last year' represented!
Insurance actuaries aggregate underwritten premiums (as well as claims and all other insurance business events…) according to inception dates (i.e. when the cover starts) and insurance accountants, according to booking dates (i.e. when the signed policy reaches the insurance firm or sometimes when it learns the policy has been accepted). Both dates can be very far apart (in some lines, 6 months isn't unusual). Hence the wildly differing sums.

These two simple examples illustrate how much individuals, teams, professions or nations can have developed contrasting practices, postulates and representations and the impact potential gaps between different worldviews can occasion.

---

[23] Alex once described actuaries as "people who found accounting too exciting, what with those yearly deadlines".

[24] "Get to the bottom of this!"

In the business world, top and middle managers daily display their preference for hard numbers, harder science and Cartesian reductionism. This is daily demonstrated:

- in planning, deciding and controlling activities,
- in financial engineering and the excesses of *management by spreadsheet*,
- in procurement practices (especially in sourcing professional services),
- in work specialization,
- in executive search.

Yet one indirect indicator after another exposes management and organizations' collective cluelessness in front of unscripted social phenomena (within which we can even include financial markets fluctuations as well as deeper crises, far removed from industrial perspectives), e.g. few executives grasp the public outrage their compensation schemes engender. Arguably, two recent events in France (the Jérôme Kerviel *vs* Société Générale '*affaire*' and the striking contrast between the despair that drove some Orange staff to suicide and top brass' attitudes), events comparable with equally symptomatic and resonant events in other countries, illustrate this further.

Perhaps this cluelessness is also reflected within these 2009 ParisTech figures:

- of its 40 privately endowed chairs, 29 leverage Social Sciences and 20 are cross-disciplinary;
- whilst, of its 177 labs, only 45 are exclusively devoted to Social Sciences and another 3 are cross-disciplinary.

**On different wavelengths? Yes. At cross purposes? Hardly.**

So populations with different life experiences, different levels or sources of education operate on different wavelengths. This is even true of management academics and management practitioners, with management consultants jockeying to bridge the gap —*tertius iungens*[25] self-serving behavior is at play.

This critical management study will therefore pore over how language may influence social representations and how different cultures and disciplines, through the refinement of their respective vocabularies, may develop contrasted symbols, of import to them, and worldviews, that make sense to them.

---

[25] Obstfeld, D. 2005. Social Networks, the Tertius Iungens Orientation, and Involvement in Innovation. *Administrative Science Quarterly*, 50(1): 100.

# II] QUESTION —ODDLY THINKING OF NUMBERS

Academia is rife with debate on many a topic. Here, neo-positivists face constructivists; there, 'quals' counter 'quants' and so on … What is striking is how little fundamental opposition exists between these points of view, how much they have in common. Indeed a prerequisite to heated debate is shared concepts on which to disagree.

The science of knowledge, epistemology, was first studied in the Western tradition by the ancient Greeks. The etymology of epistemology is derived from □πιστήμη (epistēmē), i.e. 'science, knowledge', rooted in ἐπίσταμαι (epistamai), i.e. 'I know', and -λογία (logia), 'discourse') rooted in λέγω (legō), i.e. 'I speak'.

The object of this section isn't to cover the topic of epistemology exhaustively. However, a few points will be presented in support of the approach embraced in this thesis dissertation.

**"We can know more than we can tell"[26]**

According to Nonaka[27], Polanyi[28] classified "human knowledge into two categories. 'Explicit' or codified knowledge refers to knowledge that is transmittable in formal, systematic language. On the other hand, 'tacit' knowledge has a personal quality, which makes it hard to formalize and communicate. Tacit knowledge is deeply rooted in action, commitment, and involvement in a specific context. In Polanyi's words, it 'indwells' in a comprehensive cognizance of the human mind and body." Nonaka then proposes, binarily, four modes of knowledge creation from explicit and tacit to explicit and tacit knowledge, respectively.

---

[26] Polanyi, M. 1966. *The Tacit Dimension*. London: Routledge & Kegan Paul.

[27] Nonaka, I. 1994. A dynamic theory of organizational knowledge creation. *Organization Science*, 5(1): 14.

[28] Polyani, *ibid.*

Whereas Nonaka, as well as Hedlund[29], subscribe to this view separating the natures of knowledge, Tsoukas[30] takes the more nuanced view that tacit and formal knowledge shouldn't be opposed, and that, on the contrary, they are two sides of the same coin. This is somewhat analogous to the matter-energy duality modern particle physics theory embraces. Acknowledging the dual nature of knowledge is especially important in the field of management research, as organizations are vessels of ever shifting knowledge of which a modest part is formal.

Brown and Duguid[31] offer an illustration of how, within the community of practice of copy-machine repairmen, knowledge rooted in their specific work practice is developed and shared, through such devices as narration, collaboration and social construction. The authors support their analysis of learning within this community with the viewpoint of legitimate peripheral participation first offered by Lave and Wenger[32].

The ever-shifting nature of knowledge isn't a novel concept. Heraclitus put it very aptly as "No man ever steps in the same river twice, for it's not the same river and he's not the same man."[33] Indeed, physics and chemistry tell us that at molecular and atomic levels of analysis, what we commonly tag as entities with a continuous existence may have wholly different matter within the span of a few months.

---

[29] Hedlund, G. 1994. A model of knowledge management and the N-form corporation. *Strategic Management Journal*, 15: 73.

[30] Tsoukas, H. 1996. The firm as a distributed knowledge system: A constructionist approach. *Strategic Management Journal*, 17: 11.

[31] Brown, J. S., & Duguid, P. 1991. Organizational Learning and Communities-of-Practice: Toward a Unified View of Working, Learning, and Innovation. *Organization Science*, 2(1): 40-57.

[32] Lave, J., & Wenger, E. 1991. *Situated learning: legitimate peripheral participation*. Cambridge [England] ; New York: Cambridge University Press.

[33] As quoted by Plato within Heraclitus' follower Cratylus report: Platon. 1993. *Protagoras - Euthydeme - Gorgias - Ménexène - Ménon - Cratyle*: Flammarion.

As described by Kuhn[34], academic knowledge-building is two-pronged.

Firstly, its mainstream strives to proceed with the utmost procedural formality. In support of this focus, academia ideally proceeds by method (as opposed to fiat), by logic (as opposed to faith), by conjecture and proof (as opposed to truth) and by collective (as opposed to solitary) and nominative (as opposed to anonymous) action. In the view of Nonaka (ibid.), academic knowledge proceeds to derive new explicit knowledge by combining previously known explicit knowledge.

Secondly, this gradual build-up of knowledge is occasionally shattered by a major discovery, hence the scientific revolution.

In this sense, academic knowledge, with its sudden leaps ahead following lulls, often referred to as creeps and jerks, bears a strong resemblance with evolutionary biology.

In this sense, mathematics, the ultimate 'hard', 'exact' science, is no different from organization science or any other field. It develops conceptual objects of study, posits axioms ruling these objects and extends theory through theorems, lemmas and so on. Occasionally, some of these axioms are revisited, or some implicit hypotheses and propositions are revealed leading to a new body of theory. This new body appears quite aberrant to the mainstream at first light, but sometimes becomes the mainstream to the next generation of mathematicians.

---

34 Kuhn, T. S. 1962. *The structure of scientific revolutions.* [Chicago]: University of Chicago Press.

Indeed it seems most academics hold on to the theoretical views they develop early in their career. If younger researchers challenge their views, it is not unlikely for the (revolutionary) latter to have their careers hindered by the (conservative) former. Examples[35] abound of scientific luminaries whose groundbreaking views are endorsed by their academic discipline decades after their initial publication whereas other members of the academic establishment have had their prestige wane and their contributions fade into oblivion - at least until the next shift in academic fashion.

## Mathematics, a social construction

In other words, (a) mathematics are a social construction[36]; (b) practitioners close[37] to the core of axioms involved in a particular theory have a better understanding of the assumptions made than the many more people applying mathematics.

---

[35] Two illustrations immediately come to mind:
Cantor vs Kronecker as narrated in Dehornoy, P. 2009. Cantor et les infinis. In S. M. d. France (Ed.), *Un texte, un mathématicien.*
Bachelier vs Lévy as narrated in Mandelbrot, B. B., & Hudson, R. L. 2004. *The (mis)behavior of markets: a fractal view of risk, ruin, and reward.* New York: Published by Basic Books.

[36] Berger, P. L., & Luckmann, T. 1966a. *The social construction of reality; a treatise in the sociology of knowledge* ([1st ed.). Garden City, N.Y.,: Doubleday.

[37] Though, as Kuhn (ibid.) describes, they may not wish to question the axioms in question.

Furthermore, it follows that (1) application of mathematical models occurs in consistence with Cohen et al's 'garbage can model'[38], i.e. models are adopted to make sense[39] of phenomena, not so much because of their relevance but rather because the phenomena are observed in a time frame concurrent to the mathematical model's availability; (2) early adoption by trendsetters then propagates in consistence with fashion theories[40] and late adoption by peer organization followers then propagates by mimetic isomorphism[41]; (3) widespread adopters, who are chronologically, geographically, or topologically distant from the community that set the underlying mathematical concepts will (a) have lesser understanding of these concepts' axiomatic underpinnings and (b) approach their application in a more dogmatic fashion[42].

The dominant mathematical axioms may not be truer than other, conflicting axiomatic sets, but because a broader community accepts them, they become the foundations applied wherever phenomena are modeled.

---

[38] Cohen, M. D., March, J. G., & Olsen, J. P. 1972. A Garbage Can Model of Organizational Choice. *Administrative Science Quarterly*, 17(1): 1-25.

[39] Hatch, M. J. 2006. *Organization Theory: Modern, Symbolic, and Postmodern Perspectives* (2 ed.): Oxford University Press USA, Weick, K. E. 1995. *Sensemaking in Organizations*: Sage Publications.

[40] For illustrations, see:
Abrahamson, E. 1996. Management fashion, academic fashion, and enduring truths. *Academy of Management Review*, 21(3): 616. Trepo, G. 1987. Introduction and diffusion of management tools: the example of quality circles and total quality control. *European Management Journal*, 5(4): 287. Kieser, A. 1997. Rhetoric and Myth in Management Fashion. *Organization*, 4(1): 49.

[41] DiMaggio, P. J., & Powell, W. W. 1983. The Iron Cage Revisited: Institutional Isomorphism and Collective Rationality in Organizational Fields. *American Sociological Review*, 48(2): 147-160.

[42] As Weber explains "Rationalization destroyed the authority of magical powers, but it also brought into being the machine-like regulation of bureaucracy, which ultimately challenges all systems of belief." from Weber, M. 1991. *Essays in Sociology* (1st ed.): Routledge.

The mathematical concepts this thesis will concentrate on are numbers. Numbers are learnt in primary school. This near-universality, possibly in association with the certainty in their knowledge expressed by most teachers, helps explain why few statisticians, economists, accountants, or professional managers question their mathematical toolset, a certainty researchers in 'pure' mathematics do not share[43]. Even social scientists are more likely to question their cherished axiomatic principles when they are expressed in natural language than the mathematical tools they use to shape or test their theories.

Professor Patrick Dehornoy[44], interviewed on the determination of what builds good theory (i.e. which conjectures or propositions are elevated to axiomatic status), offers the following musing[45]:

---

[43] Once, during a math lesson at Ecole Polytechnique conducted by Professor Laurent Schwartz, an eager *polytechnicien* suggested: *« On pourrait employer ça dans* (ce contexte). *»* ["We could use this in (this context)."] Prof. Schwartz swung around violently and uttered *« Mais nous sommes purs ! »* ["But we are pure!]

[44] Dehornoy, P. 2009. Cantor et les infinis. In S. M. d. France (Ed.), *Un texte, un mathématicien.*

[45] In essence: "That is the question. (…) The question you raise is 'what's a good axiom?' (…) I don't know if there is a right answer. (…) When axioms produce a description of the world of mathematics that seems satisfactory, little by little, that's when there's a consensus, but it's a dynamic and slow phenomenon. … Take the existence of infinity: for Kronecker, infinity could not exist. For all of you, today, a century later, at last, there is no one in mathematics or nearly no one, rejecting the axiom of infinity. … I sometimes ask myself what Martian math are like. Perhaps they have no infinity. (…) We've lived with infinity since high school, so it's natural to us. (…) Today, in the small, small world of set theorists, transfinite cardinals are as natural as infinity, (…) are as concrete; their truth is evident to us for the moment. But this is not yet shared by everyone. (…) It's a consensus within a community. It's not dictated by hard laws. Little by little, with hard work, over fifty years, by tens of mathematicians, a world view emerges that seems good, that seems satisfactory … And perhaps, certainly, other axioms, beyond what we have in the 20th century, will become self-evident too."

« *C'est la question. Evidemment, quand on a un système incomplet, on peut toujours se dire ‹ je rajoute quelque chose ; je prends l'hypothèse du continu parce que je l'aime bien et la rajoute comme axiome ; du coup, si je la mets comme axiome, elle sera vraie ›. La question que vous soulevez est ‹ qu'est-ce que c'est qu'on bon axiome ? comment choisit-on les axiomes ? › Et là, c'est là que c'est important. Ma réponse, la réponse – je ne sais pas s'il y a une bonne réponse – la réponse que je serais tenté de faire est la suivante :*

*On ne choisit pas un axiome par une réflexion a priori. On ne s'assoit pas dans le noir à une table en se disant ‹ Oh la la qu'est-ce que je pourrais rajouter comme axiome qui ferait malin ? › Ce n'est pas ça. On explore des tonnes d'axiomes, des tonnes, des tonnes, des tonnes, mais vraiment des tonnes d'axiomes possibles et puis on voit que certains donnent une description du monde mathématique qui est satisfaisante et d'autres qui donnent une description du monde mathématique qui semble complètement pathologique.*

*Et quand il y a des axiomes qui donnent une description qui semble satisfaisante, petit à petit, c'est là qu'il y a un consensus mais c'est un phénomène dynamique et lent. Au bout de dix ans, au bout de quinze ans, il y a un consensus qui émerge pour dire ‹ ah ben oui, la description du monde mathématique donné par cet axiome est correcte ›.*

*C'est exactement la même chose que l'existence de l'infini. Pour Kronecker, l'existence de l'infini n'est pas possible. Pour chacun de vous, aujourd'hui, cent ans après, finalement, il n'y a plus personne aujourd'hui en mathématiques ou pratiquement personne pour refuser l'axiome de l'infini (il existe des infinis). C'est devenu intuitif. Tout le monde trouve que c'est naturel. Pourquoi ? Parce que depuis votre petite enfance, depuis le lycée, depuis toujours, vous manipulez l'infini, les limites de réels, etc. et du coup, c'est devenu familier.*

*Donc ma vision des choses est que l'évidence n'est pas quelque chose d'inné, de donné une fois pour toutes mais plutôt l'intériorisation d'une longue familiarité. Vous êtes familiers avec l'infini et du coup, ça devient tellement naturel que vous finissez par trouver évident l'axiome de l'infini.*

*Je vais vous dire un truc qui me fait toujours beaucoup rêver : je me demande ce que sont les mathématiques des martiens. Sur la planète Mars, il y a peut-être des mathématiques. Il se pourrait très bien qu'il y ait une planète où il y ait des mathématiques sans l'infini. Et finalement, ça ressemblerait beaucoup à nos mathématiques ; ils auraient le même théorème de Fermat, toute la théories des nombres, l'analyse, ça ressemblerait beaucoup mais ils n'auraient pas l'infini. Mais nous sommes familiers avec l'infini depuis un bout de temps et actuellement on le trouve naturel.*

*Ce que je vous disais à propos des axiomes de grands cardinaux, aujourd'hui, dans le petit monde, petit, des théoriciens des ensembles, les grands cardinaux, on y croit, ça nous paraît tout aussi naturel que l'infini. Les grands cardinaux ne sont jamais que le grand infini, en fait ; c'est tout pareil. Et pour nous aujourd'hui, c'est aussi concret, aussi naturel donc pour nous, ça nous paraît évident que c'est vrai mais pour le moment, ce n'est pas encore partagé par tout le monde parce que les gens n'en ont pas entendu parler. C'est plutôt ça. C'est un consensus dans une communauté. Ce n'est pas dicté par – ce ne sont pas des lois d'airain. Personne n'a gravi le mont Sinaï écouter Dieu leur dicter son axiome. C'est petit à petit, à force de travail, de cinquante ans de travaux de dizaines de mathématiciens qu'on voit émerger une vision du monde qui semble bonne, qui semble satisfaisante.*

*Et du coup, sur cette base, ZF (la théorie des ensembles de Zermelo-Fraenkel), c'est ça. Zermelo a proposé une vision du monde des ensembles et les gens ont dit ‹ ben oui c'est raisonnable ; on est d'accord ; OK on y va ›. Peut-être les martiens n'ont-ils pas du tout cette vision-là. Mais, nous, les humains, on partage ça. Et ce que je dis est qu'un jour, il y a peut-être d'autres axiomes, il y a certainement d'autres axiomes au-delà de ce qu'on a au XX^e^ siècle qui susciteront la même évidence. Nos petits-enfants diront ‹ c'est évident, les grands cardinaux, bien sûr ! Mon papa est un grand cardinal ! › »*

French *académicien des sciences* Jean-Pierre Ramis in another interview[46] reports that divergent series, first studied by Euler[47], is a mathematical topic little studied by the mathematics community until the late 1980s, whereas theoretical physicists had found potential in divergent series much earlier to model phenomena.

## Mathematics, a technology borne by minds

Indeed, each mathematical set of coherent paradigms may be considered to be a technology and each shift, a technological evolution.

Judgments about the legitimacy of mathematical theories are no different from other judgments about legitimacy on other matters. As Granovetter[48] notes, "Judgments about legitimacy are part of larger normative frameworks that people produce in all known social structures. Norms do not come from above, nor do they arise in most cases from some evolutionary process that selects for efficiency. Instead, they are enacted, reproduced, or changed in the course of each group's everyday social activity. Norms may be in part a reflex of changes in practical realities, as when increasing women's participation in the workforce makes sexist conceptions appear quaint and outdated. Yet they have some life of their own and can predate and influence as well as emerge from social change." Adoption of mathematical concepts will thus be legitimized by its repeated enactment and may both emerge from and influence social change.

---

[46] Ramis, J.-P. 2009. Leonhard Euler, ou l'art de donner un sens à ce qui n'en avait pas. In S. M. d. France (Ed.), *Un texte, un mathématicien.*

[47] Euler, L. 1746. De seriebus divergentibus. *Opera Omnia*, I(14): 585-617.

[48] Granovetter, M. 2007. The Social Construction of Corruption. In V. Nee, & R. Swedberg (Eds.), *On Capitalism*: 152-172: Stanford University Press.

Norms are hard to change, and groups in conflict will advance different narratives to promote their conflicting interests. As Ewick and Silbey[49] suggest, groups that can subsequently impose a hegemonic discourse that shapes peoples' understanding gain a powerful advantage over others.

Sociological, philosophical, and economics theories, as explicit knowledge, are epistemologically as codified as mathematical theories. All deserve measures of skepticism as to their foundation, their validity, and their role in shaping reality as socially represented.

In other words, (a) mathematics are a social construction[50]; (b) practitioners close to the core of axioms involved in a particular theory understand better the assumptions made that the many more people applying mathematics; (c) mathematics are sometimes applied in situations akin to the "garbage can[51]" model, as observation of phenomena and availability of the mathematical model occur in a concurrent time frame; (d) models are adopted to make sense of phenomena[52], not so much because of their relevance but rather (e) applying mathematics in an organizational context may lead to their mimetic application in peer organizations[53]; (f) adoption by trendsetters then propagates in consistence with fashion theories[54]; (g) widespread adopters, chronologically, geographically, or topologically distant from the community that set the underlying mathematical concepts have a lesser understanding of these concepts' axiomatic underpinnings and approach their application with the machine-like regulation of bureaucracy[55]; (h) mathematical concepts will be legitimized by repeated enactment, and social change may both emerge from and influence their adoption[56]; (i) groups that succeed in imposing their discourse shape peoples' understanding and gain a powerful advantage over groups with which they are in normative conflict[57].

---

[49] Ewick, P., & Silbey, S. S. 1995. Subversive Stories and Hegemonic Tales: Toward a Sociology of Narrative. *Law & Society Review*, 29(2): 197.

[50] Berger, P. L., & Luckmann, T. 1966b. *The Social Construction of Reality: A Treatise in the Sociology of Knowledge*: Anchor Books. Piaget, J., & Szeminska, A. 1941. *La genèse du nombre*

Maslow[58] aptly put it thus: "I suppose it is tempting, if the only tool you have is a hammer, to treat everything as if it were a nail". Few statisticians, economists, accountants or professional managers behaviorally resist the temptation to apply mathematics as a solution looking for a problem.

## On the social grounding of technology ruptures

Whenever a new technology is introduced into a society (e.g. sewers, steam engines, railways, electricity, standardized containers, the Internets…) new professions are generated.

---

*chez l'enfant.* Neuchâtel: Delachaux & Niestlé.

51 Cohen, M. D., March, J. G., & Olsen, J. P. 1972. A Garbage Can Model of Organizational Choice. *Administrative Science Quarterly*, 17(1): 1-25.

52 Weick, K. E., & Bougon, M. 1986. Organizations as cognitive maps: charting ways to success and failures. In H. P. Sims, Jr., & D. A. Gioia (Eds.), *The thinking organization*: 102-135. San Francisco: Jossey-Bass.

53 DiMaggio, P. J., & Powell, W. W. 1983. The Iron Cage Revisited: Institutional Isomorphism and Collective Rationality in Organizational Fields. *American Sociological Review*, 48(2): 147-160.

54 Abrahamson, E. 1996. Management fashion, academic fashion, and enduring truths. *Academy of Management Review*, 21(3): 616.

Kieser, A. 1997. Rhetoric and Myth in Management Fashion. *Organization*, 4(1): 49.

Trepo, G. 1987. Introduction and diffusion of management tools: the example of quality circles and total quality control. *European Management Journal*, 5(4): 287.

55 Weber, M. 1991. *Essays in Sociology* (1st ed.): Routledge.

56 Granovetter, M. 2007. The Social Construction of Corruption. In V. Nee, & R. Swedberg (Eds.), *On Capitalism*: 152-172: Stanford University Press.

57 Ewick, P., & Silbey, S. S. 1995. Subversive Stories and Hegemonic Tales: Toward a Sociology of Narrative. *Law & Society Review*, 29(2): 197.

58 Maslow, A. H. 1966. *The Psychology of Science: A Reconnaissance.* New York: Harper & Row.

Since Antiquity, geography has structured the material economy. The cost of materials transformation depends on proximity to the raw materials to transform, the optimal combination of competitively priced resources, the ease of logistical access and the availability on site of a competent workforce. In past civilizations, every particular instance of materials transformation usually ended either with exhaustion of the raw materials from the site[59] or with a technological rupture.

Early civilizations are autarchic. Supply and demand are in spatial equilibrium, or rather in local temporal flux. As exchanges developed, the location of demand becomes just as important, i.e. buying power attracted suppliers.

Each era's economic activity reflects its technological idiosyncrasies. Technological change has driven economic development since agriculture displaced hunting and gathering. Later, civilizations were borne by their mastery of a particular generation of steel transformation, sewage systems…

Each new technology disrupts previous lifestyles and production processes, imposes its own specificities, introduces its own network of organizations and economic agents (language/media/communication, education/knowledge, water, energy, transport…), which represent large commitments to infrastructure and introduce new professions. These new professions act as Temple guardians, moved by a mixture of (real or perceived) safety and quality imperatives as well as by their collective desire to perpetuate their guild.

The breadth (local, regional, national, global) of these networks and their economic counterparts is also specific to each era.

Indeed, accounting practices for the various trades and the manner in which goods and services are categorized reflect the era in which they occur.

---

[59] *e.g.* How trees became extinct on Easter Island.

What is commonly referred to as globalization describes the process by which scaling effects, new sourcing strategies, and capability concentration have redrawn economic activity in recent times.

These transformations, however, do not result in the optimal reshuffle of activity the proponents of utility-function-driven microeconomic models have put forward. Two notable studies aptly illustrate the social emergence of technology.

In their study of the origins and early development of the electricity industry in the United States, Granovetter and McGuire show[60] this industry was "born not of Benthamite Equations or optimizing rationality, but longstanding friendships, similar experiences, common dependencies, corporate interlocks, and creation of social relations. Samuel Insull[61] and his circle of collaborators socially constructed their firms in similar ways, and then promoted a system of industry governance and template diffusion. They drew upon their national contacts to reframe the market and the political system in ways that pressured utility firms toward technical, organizational, economic, and legal conformity. Yet, isomorphism among firms was never fully achieved because this was a system of influence and not of direct control, and because of the varying resources and market attributes in each firm's locale.

---

[60] Granovetter, M., & McGuire, P. 1998. The Making of an Industry: Electricity in the United States. In M. Callon (Ed.), *The Laws of The Markets*: 147-173. Oxford: Blackwell.

[61] A close associate of Thomas Edison, credited with co-founding what is now General Electric.

Such a concentrated pattern of friendship, shared experience, club activity, and domination through an industry governance structure in the process of industry formation does not necessarily continue indefinitely... Were we to follow the industry past its formative years, we would find that processes and relationships once shaped by individuals became institutionalized in more formal organizations, institutional alliances, standardized practices, and industry norms. As in other industries, such patterns become embedded as norms, unless and until an industry crisis occurs. At that point, a new social network of firm leaders has the potential to redefine and reconstruct a new industry based upon various elements of the old."

Individuals mobilizing financial, technical, and political resources through their social and professional networks pushed the industry in certain directions, with substantial historical contingency and path dependence. Technology and organizational forms were not simply responses to technical or economic exigencies; in fact, technical and organizational forms shunted aside in this period, and then forgotten, were far more plausible technically and economically than usually supposed. Some such forms, such as the decoupling of distribution from production or on-site power generation in homes and factories, began to reappear in the late 20th century.

Less recently, Cowen argues[62] technology can best be understood as the product of social preferences, bureaucratic mandates, and economic pressures. Her seminal study of how the "refrigerator got its hum" shows that existing (quiet) gas refrigerators were not a casualty of some technical necessity. Rather, General Electric in the 1920s and 1930s mounted a huge effort to develop and sell to U.S. households electric refrigerators that just coincidentally required an ideal 24-hour-a-day electrical supply from the power utility, General Electric.

---

[62] Cowen, R. S. 1985. How the Refrigerator Got Its Hum. In D. MacKenzie, & J. Wajcman (Eds.), *The Social Shaping of Technology*: 208-211. Buckingham: Open University Press.

That these two studies reflect GE's exercise of its social power, despite on-paper superior electrical technology in the first example, and more consumer-friendly non-electrical technology in the second, reinforces the case against technological determinism and laissez-faire market forces.

The underlying structural theory of power driving technological choices can be traced to Marx[63]. Marx's concern is with how powerful groups manage to get their way, using force and persuasion to promote the management practices and public policies they favor. These groups succeed not by force alone, but by promoting what Marxist Georg Lukács[64] calls "false consciousness" to convince people that what is in these groups' interest is actually in the common (corporate, national, or global) interest.

## On digital technologies

Digital technology disruptions share the same dynamic as previous disruptions but their spread, contrary to previous disruptions, is unhindered by logistical obstacles and only mildly constrained by material geography[65]. On the other hand, practices derived from technological disruptions spread according to cognitive topologies. Today, digital technologies are widespread as they support tasks in our personal lives as well as in agricultural, industrial, and service processes.

As computing automates administrative processes, transforms customer support with self-service devices, or automates production, a shift in the value chain[66] towards digital resources is under way. This mutation, however, is more fundamental. Not only are existing products, services and organizations transformed, new digitally supported products, services, and organizations are made possible.

---

63 Marx, K. 1894. *Das Kapital*. New York: International Publishers.

64 Lukacs, G. 1920. *History & Class Consciousness* (R. Livingstone, Trans.): Merlin Press.

65 Bandwidth and latency issues still play a role resp. in high volume, high speed contexts that explain clustering, virtualization and replication mechanisms.

66 "transfert de charge".

Organizations are distinguished by the services they offer, their know-how, and their capabilities, including their capability to leverage information and communication technologies. The information system of any organization supports the services it offers and underlying commercial, financial, and administrative exchanges.

Organizations are transformed both internally and externally by the digital mutation. This mutation is especially disruptive due to the confluent emergence of networked, pervasive terminals and grid, cloud computing. The introduction of these has shattered traditional conventions of time and space (as they relate to work/life balance, to team building, even to identity…) in ways not experienced since the dawn of the industrial revolution, which saw a novel concentration in activity in electrically powered factories. In fact, the digital mutation may be the most significant disruption since agriculture displaced hunting and gathering.

Internally, information and communication technologies strive to make the supply chain, inventory management, logistics, quality control, and audit trails leaner, more flexible, or more adaptable. Moreover, traditional administrative, bureaucratic structures, such as those found notably in public administrations and financial services, have automated much of their information processing.

Externally, roles and boundaries between organizations are shifting. Outsourcing has grown, spurred by the apparent lowering of agency costs.

Every industrial sector has seen such a wane in the relative importance of materials-based economic activity that national policies are struggling to cope with the shift. It is true that some activities, such as banking and insurance, were always preeminently immaterial.

In a sense, language, knowledge, and mathematics already form a millennial technology. The manner in which language, knowledge, and mathematics disseminate within transactive memory systems[67] prefigures the logistical contrast between traditional materials-based activities and the digital activities the advent of electronics and information and communication technologies have enabled.

Spatial and temporal distances have been tumbling. Industries are being redefined and technologies such as the Internet are playing a role in the shuffling of industrial activities and capabilities.

## On Information and its categorization

Aristotle[68], in the first section of his Organon, organizes information attributes into four separate kinds: (a) enumerated information, (b) spatial information, (c) temporal information, and (d) coded information.

Enumerated information is information representing a quantity of something being counted and whose elements are indistinguishable from one another (i.e. 1, 2, or 31). A measure of length, weight, … is not enumerated information, as it requires a comparison with a reference or standard. In Aristotle's view, such information is coded. Similarly, negative numbers (e.g. -3), percentages (e.g. 42%), fractions (e.g. ¾), algebraic numbers (e.g. √2), and geometric numbers (e.g. π) are objects of convention with no tangible reality, i.e. socially constructed abstractions.

Spatial information conveys a physical reality (i.e. here, there, where we are). It is always subjective, though its subjectivity may be collectively shared.

---

67 Peltokorpi, V. 2008. Organizational Transactive Memory Systems: Review and Extension.

68 Aristotle. 2000a. *Catégories de l'interprétation: Organon I et II*: Vrin.

Temporal information conveys a temporal reality (i.e. now, before, soon). It too is always subjective, though its subjectivity may be collectively shared.

Coded information is information coded according to a pre-set convention (i.e. felines, organizational justice, the USA, in the Pyreneans, on Alpha Centauri, the number 0, November 22nd 1963, or 15 lbs.). This is the most common type of information and as the existence of a pre-set convention implies, it is the most blatantly socially constructed of the four kinds of information attributes earmarked by Aristotle.

All four kinds are described in Table 1.

| **Kind of information attributes** | **Definition** | **(E) Explicit and (I) implicit examples** | **Comments** |
| --- | --- | --- | --- |
| Enumerated | Quantity counted of elements | E: 1, 2, 31<br>I: a donkey | universal; applies to undistinguished elements |
| Temporal | Reference to temporal realities | E: now, before<br>I: He will grow up | universal, albeit relative; relativity may be shared between enunciator and correspondents |
| Spatial | Reference to physical realities | E: there, where we are<br>I: Come! | |
| Coded | Reference to pre-set convention | E: in the Andes, East Germany, 0, 2/7/1828, OB, felines, Socrates, ¾, on Orion | most common category; most blatantly socially constructed |

*Table 1: Kinds of information attributes derived from Aristotle*[69]

[69] Aristotle (*ibid.*)

Information is sometimes implicitly carried. Use of the future tense is an example of implicitly carried temporal information.

So it seems (a) enumerations represent the only objectively defined category; (b) spatial information merely requires understanding where the speaker spoke said spatial information; and (c) temporal information, understanding when the speaker spoke said temporal information.

All other forms of information require a shared context, beyond a common language, to be understood.

So, in Western tradition harking back to Greek Antiquity, numbers, or more specifically integers, represent the only seemingly universal and absolute type of information.

## On the pantometric[70] *mentalité* of modern Occident

Crosby[71] proposes that Western civilization emerged as the dominant culture of the contemporary world due to the dual accomplishment of mastering quantification and visualization techniques during the period spanning the mid-13$^{th}$ to 16$^{th}$ centuries.

These changes, brought about by emphasis upon visualization and measurement, are traced through the concepts of time, space, and mathematics as they developed in Western experience during the 1250-1600 CE period. Making the first clocks to measure time, constructing buildings that were measured by uniform quanta to achieve effective proportions, and ordering numbers on something other than an abacus device turn out to be crucial fundamental steps. This is what enabled the West to succeed in controlling the environment as no other culture has.

Crosby describes "Westerners (as) monotheists fascinated with light, gloried in pantometry".

---

70 Pantometry, a term coined by Crosby, describes the activity of universal measurement (i.e. consists in measuring everything).

71 Crosby, A. W. 1997. *The Measure of Reality: Quantification and Western Society, 1250-1600*: Cambridge University Press.

The contemporary controversy about empirical quantification and qualitative research can be traced to how quantification changed Western civilization.

The pantometric *mentalité* that pervades contemporary thought to this day is epitomized by Thorndike[72]'s oft-quoted "Whatever exists at all exists in some amount. To know it thoroughly involves knowing its quantity as well as its quality" and, closer to our field, Kuhn[73]'s "Numbers register the departure from theory with an authority and finesse that no qualitative technique can duplicate, and that departure is often enough to start a search". More recently, Mangin[74], in his breakdown of performance management systems design aptly notes "Decisions are about premises, since logical and mathematical deduction leaves no room for human discretion: once the problem is (politically) well defined, it is (logically) already solved."

## On language and meaning

Indeed, assuming one wishes to communicate information to another, if this information is code (as in the case of 12 lbs, Paris or †), that code must be shared; if it is spatial, temporal or an enumeration (as in the case of here, soon or 12), no more than the language need be common. On the other hand, spatial and temporal will be understood as relative to the orator and the context of his oratory whereas, under our present conditioning, enumerations will commonly be regarded as universally absolute to all but the hardcore mathematics community.

---

72 Thorndike, E. L. 1918. The nature, purposes, and general methods of measurements of educational products. In G. M. Whipple (Ed.), *The Seventeenth yearbook of the National Society for Study of Education. Part II. The Measurement of Educational Products*: 16. Bloomington, IL: Public School Publishing Co.

73 Kuhn, T. S. 1961. The Function of Measurement in Modern Physical Science. *Isis*, 52(2): 161-193.

74 Mangin, N. 2008. A comprehensive analytical framework for studying performance evaluation, *2008 Academy of Management Annual Meeting*. Anaheim, CA: Academy of Management.

## Theoretical propositions

I put the following propositions forward, drawn from Aristotle's breakdown of language elements into four types (and previously described in Table 1).

### The enumeration proposition

> **P1** Western thought and semantics prime thinkers and speakers into attributing an a priori universal and absolute quality to enumerations.

Enumerations —i.e. non-null integers— play a special role in English. If two people speak the same language and the former's discourse contains an enumeration, e.g. 58, the latter needs nothing more to understand its meaning.

### The spatial proposition

> **P2** Western thought and semantics prime thinkers and speakers into attributing an a priori universal and relative quality to spatial information.

Spatial information —i.e. here, there, yonder— plays a special role in English too. If two people speak the same language and the former's discourse contains a spatial indication, e.g. above, the latter needs nothing more to understand this indication in question than the spatial frame of reference the speaker implicitly or explicitly is alluding to.

### The temporal proposition

> **P3** Western thought and semantics prime thinkers and speakers into attributing an a priori universal and relative quality to temporal information.

Temporal information —i.e. soon, now, long since— plays a special role in English similarly to spatial information. If two people speak the same language the former's discourse's temporal contents, e.g. the tense of a verb, give the latter enough to understand these contents in question than the spatial frame of reference the speaker implicitly or explicitly is alluding to.

## The social code proposition

> **P4** Western thought and semantics prime thinkers and speakers into attributing an a priori subjective and shared, i.e. social quality to coded information.

This proposition is put forward merely to exhaustively encompass Aristotle's breakdown of language elements. No attempt will be made to refute this already well-established proposition.

## Undo the math !

A better approach to patching is to look for unifying principles that allow a single general component to replace a set of narrow, specialized components.

However, Thorngate[75]'s (im)postulate of commensurate complexity states: "a theory of social behavior cannot be simultaneously general, accurate, and simple; two of the three characteristics are possible, but only at a loss to the third." Oddly, as Weick[76] reports, Thorngate's impostulate doesn't seem to apply to itself.

---

[75] Thorngate, W. 1976. "In General" vs. "It Depends": Some Comments of the Gergen-Schlenker Debate. *Personality and Social Psychology Bulletin*, 2: 404-410.

[76] Weick, K. E. 1999. Theory construction as disciplined reflexivity: tradeoffs in the 90s. *Academy of Management Review*, 24(4): 797-806.

What one can infer from these musings is that the reflexive nature of social behavior warrants theories supported by finitely recursive and/or reflexive postulates, i.e. theories, as a whole or in parts, must incorporate themselves and mirror this self-incorporation in a finitely recursive fashion.

# III] PROTOCOL —BEYOND KNOWN SPACE, TIME, NUMBERS

« *Je serais enclin à dénier la qualité proprement philosophique à toute œuvre où ne se laisse pas discerner ce que j'appellerai la morsure du réel. J'ajouterai que, dans le développement d'une philosophie, il y a malheureusement presque toujours un point à partir duquel l'instrument dialectique tend à jouer tout seul, c'est-à-dire à vide.*[77] »
Gabriel Marcel[78].

### Social networks theory

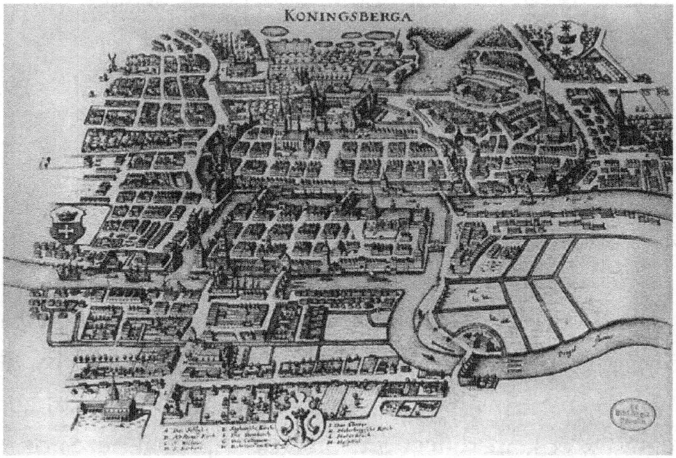


*Figure 2: Konigsberg and its 7 bridges*

Historical streams originate with Leonhard Euler[79] (whose paper on the Konigsberg Bridge problem is recognized as the seminal paper on topology and graph theory) on the one hand and Georg Simmel (who is recognized for introducing the concept of social structure) on the other.

---

[77] "I would be inclined to deny the title of philosophical to any work which fails to reflect what I will call the bite of reality. I will add that, in the development of a philosophy, unfortunately there is almost always some point at which the dialectical device tends to run by itself; that is, spinning its wheels."

[78] Marcel, G. 2005. *Tu ne mourras pas.*

[79] Euler, L. 1741. Solutio problematis ad geometriam situs pertinentis. *Commentarii*

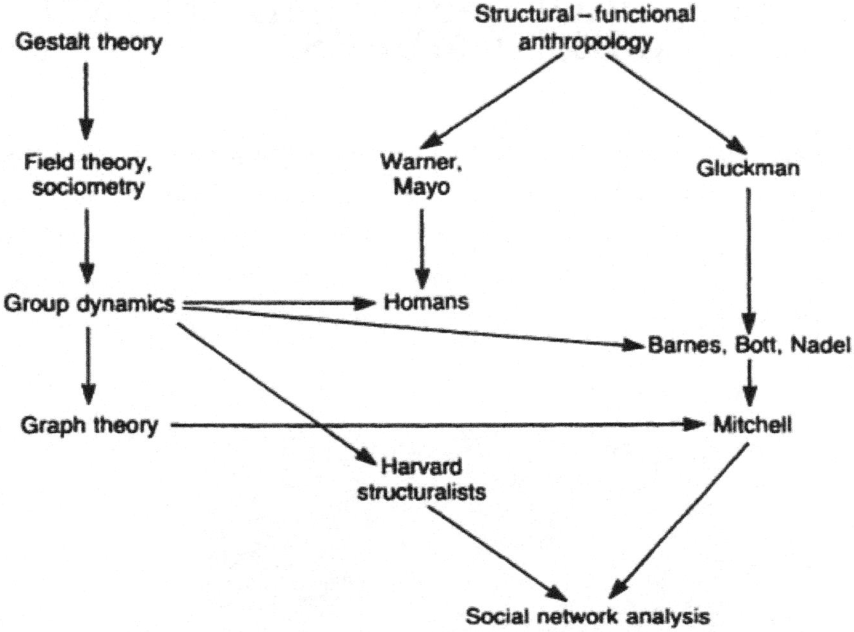


*Figure 3: Pre-computational origins of social networks analysis*[80]

Prior to this, Plato[81] had decided that the number of citizens in the ideal state was 5040. However, Plato chose it because it is divisible by all numbers up to 12 (bar 11). So his reasoning was more mathematical and abstract than integrative of a reflection on the optimal size of a community in relation to a manageable number of interpersonal ties. The Dunbar number, a theoretical cognitive limit to the number of people with whom one can maintain stable social relationships, has on the other hand been estimated to be between 150[82] and 290 with a median of 230[83].

---

*academiae scientiarum imperialis Petropolitanae* (8): 128-140.

80 Scott, J. 2000. *Social network analysis - A handbook* (2nd ed.): Sage Publications.

81 Platon. 1831. *Nomoi* (V. Cousin, Trans.): Pichon.

82 Dunbar, R. I. M. 1993. Coevolution of neocortical size, group size and language in humans. *Behavioral and Brain Sciences*, 16(4): 681-735.

83 Bernard, H. R., Shelley, G. A., & Killworth, P. D. 1987. How much of a network does the GSS and RSW dredge up? *Social Networks* (9): 49-61.

Bernard, H. R. 2006. Honoring Peter Killworth's contribution to social network theory.

Whilst not setting his sights on any given threshold, Simmel[84], very mathematically suggests there must be an upper limit to the close circle one can constitute. He defines this circle as a tight group sharing a common set of values or goals – in his words a positive definition of a social group – which he contrasts with the community at large within which it circumscribes itself. He defines this community as a loose group merely sharing a few prohibitions – a negative definition of a social group. He also notes that between the One –the individual– and the All –the nation–, German society divided into Hundreds, an administrative segmentation that kept its name even when subgroup figures fluctuated, all the more to reflect the impersonal nature of the Hundred.

---

University of Southampton.

84 Simmel , G. 1902. Georg Simmel. "The Number of Members as Determining the Sociological Form of the Group: I. *The American Journal of Sociology*, 8: 1-46.

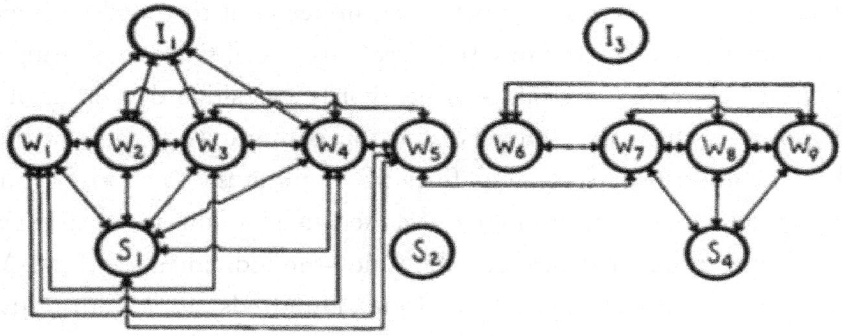


*Figure 4: Hawthorne Works study Participation in Games diagram*

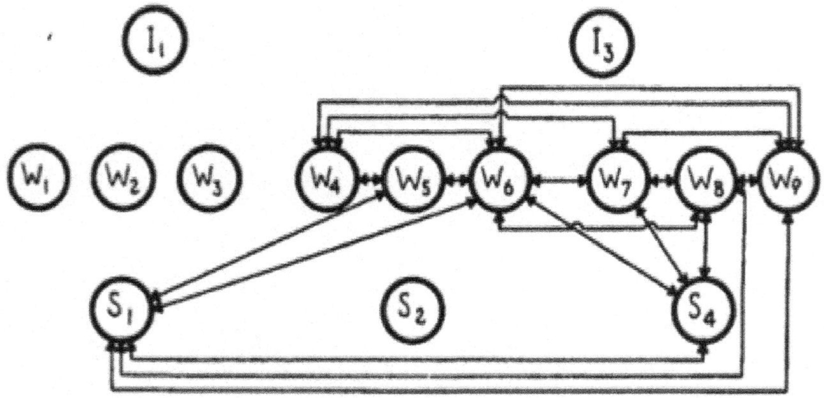


*Figure 5: Hawthorne Works study Participation in Controversies about Windows diagram*

Roethlisberger, Dickson and Mayo conducted an early illustration of social networks analysis during the seminal Hawthorne works study[85].

Another notable early expert was Moreno[86]; one of his famous sociograms is reproduced in Figure 6.

---

[85] Roethlisberger, F. J., Dickson, W. J., & Wright, H. A. 1941. *Management and the Worker: An Account of a Research Program Conducted by the Western Electric Company, Hawthorne Works*: Harvard University Press.

[86] Moreno, J. L. M. D. 1953. *Who Shall Survive?* (2nd ed.).

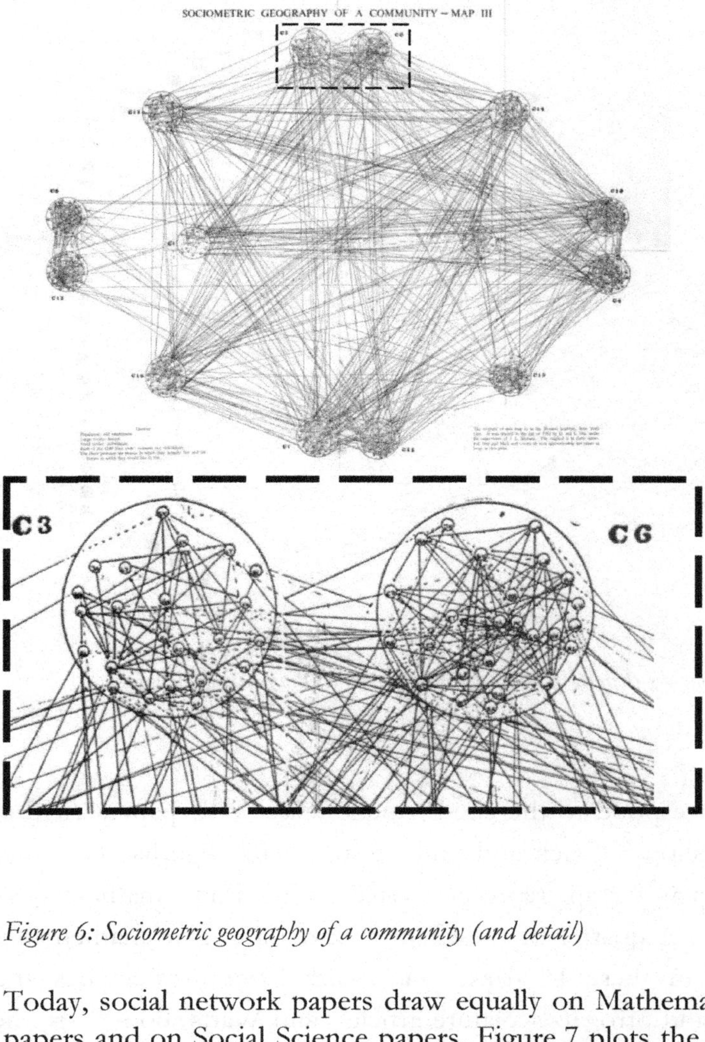


*Figure 6: Sociometric geography of a community (and detail)*

Today, social network papers draw equally on Mathematics and Physics papers and on Social Science papers. Figure 7 plots the citation patterns of social network papers published in 2005 in the *American Journal of Sociology* and the *American Sociological Review*.

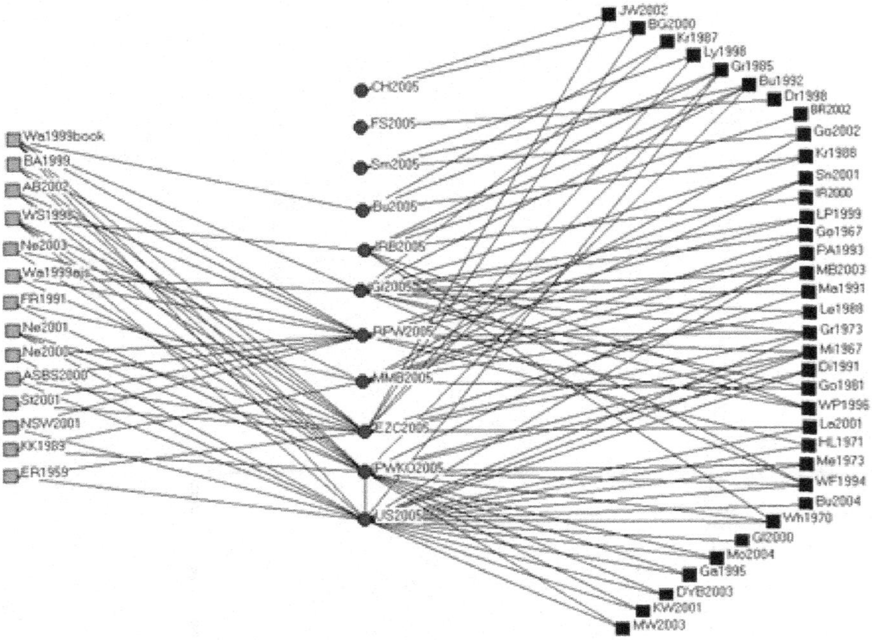


*Figure 7: Social networks-themed AJS and ASR papers citation network*[87]

The squares correspond to the cited papers, where only papers cited at least twice are shown. Black and pink (resp. on the righthandside and lefthandside) squares resp. represent social sciences and mathematics/ physics research. A quarter of the citations listed are from mathematics and physics. 8 of these 11 papers cite work from mathematics and physics. Watts and Strogatz's Nature article[88] and Watts' book[89] are the most cited. They have built their structural analysis of the small world phenomenon on Erdős and Rényi[90]'s seminal paper defining contemporary graph theory.

---

[87] Lazer, D., Mergel, I., & Friedman, A. Co-citation of prominent social network articles in social science journals: Showing the change in research paradigm - how sociologists include references to science articles in their work.

[88] Fligstein, N. 1996. Markets as politics: A political-cultural approach to market institutions. *American Sociological Review*, 61(4): 656, Watts, D. J., & Strogatz, S. H. 1998.

## Social Networks *vis-à-vis* the Hierarchy vs Market debate

Many an academic has weighed in on the hierarchy vs market debate[91]. Suffice it to say that some contexts favor the emergence of hierarchies, others markets or, rather, at the level of analysis (individuals, firms, industries, nations) relevant to the context, a first approximation in understanding the given context is either (a) consistent with a hierarchal model with one or more trees of actors or (b) consistent with a total lack of hierarchal relationships (akin to disorder, entropy) with all actors free agents, level with one another, i.e. the market model.

Moullet[92] observes that (a) within real apparent hierarchies, the posted rules, procedures and formal details mask the negotiations, trades and contracts that led to them behind closed doors and (b) conversely, within real apparent markets, the tangible negotiations, trades and contracts conceal the rules, procedures and formal details that shaped them.

The underlying truth is that man is a relational animal, that spatial and temporal proximity as well as cognitive propinquity deepen and broaden human relationships and these gradually built entwined social relationships help explain many interpersonal behavioral patterns, individually and on an aggregate level.

---

Collective dynamics of small-world networks, *Nature*, Vol. 39: 440-442.

89 Watts, D. J. 1999. *Small Worlds*: Princeton Univ. Press.

90 Erdős, P., & Rényi, A. 1959. On Random Graphs. *Publicationes Mathematicae Debrecen*, 6: 290-297.

91 To illustrate a few contrasting approaches:

Aoki, M. 2004. An organizational architecture of T-form: Silicon Valley clustering and its institutional coherence. *Industrial and Corporate Change*, 13(6): 967.

Fligstein, N. 1996. Markets as politics: A political-cultural approach to market institutions. *American Sociological Review*, 61(4): 656.

Williamson, O. E. 1995. Hierarchies, Markets and Power in the Economy: An Economic Perspective. *Industrial and Corporate Change*, 4(1): 21.

92 Moullet, M. 1992. *Le management clandestin*: Dunod.

Simmel[93] notes "this of course means, on the one hand, not the abstract consciousness of the unity concept, but the innumerable singular relationships, the feeling and knowing about this determining and being determined by the other, and, on the other hand, it quite as little excludes an observing third party from performing in addition a synthesis, with its basis only in himself, between the persons concerned, as between special elements. Whatever be the tract of externally observable being which is to be comprehended as a unity. The consummation occurs not merely by virtue of its immediate and strictly objective content, but it is determined by the categories of the mind (Subject) and from its cognitive requirements. Society, however, is the objective unity which has no need of the observer not contained in itself."

In short, (a) the hierarchal model represents the engineer's ideal of an organizational machine that is no more than the sum of its parts, usually centrally controlled; (b) the market model treats every node as equal and independent, an economist's ideal inspired by classical physics, with everything from sub-atomic particles to heavenly bodies influencing each other in a deterministically Pickwickian fashion.

But social organizations are more than the sum of their parts and one's influences on anyone else won't comply with a simple formula independent of the one, the other and the surrounding circumstances and history of the one, the others and their surroundings.

Social networks are the links we weave and unweave by which immaterial notions such as information, trust, affection… ebb and flow between people (and more broadly sentient beings).

93 Simmel, G. 1910-11. How is society possible? *The American Journal of Sociology* (16): 372-391.

A hierarchal organization may be defined and described objectively. A market need only be defined by enumerating its agents and the nature of their theoretically neutral interactions. A social network is innately subjective, as distributed and flowing in nature as tacit knowledge. Whenever one is reported, its point or points of reference (i.e. a single or many observers) must be explicitly expressed. Usually, a social network chart is described graphically in a fashion similar to Figure 8.

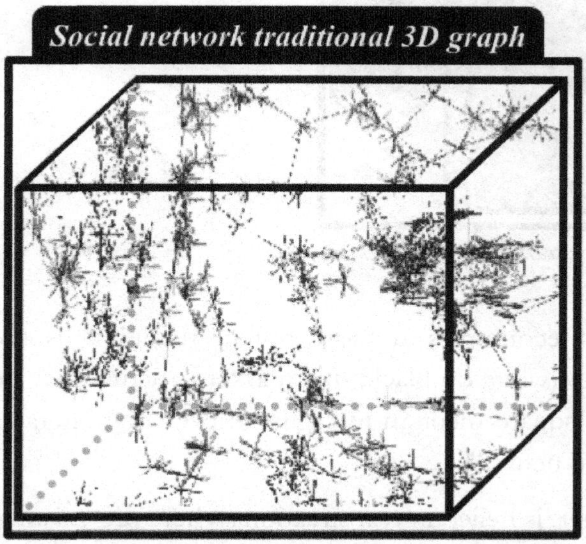


*Figure 8: A traditional, three-dimensional graphical rendering of a social network*

This typical black on white rendering is inherited from organization charts of hierarchies. It reflects the grounding of social network theory in organization studies and sociology.

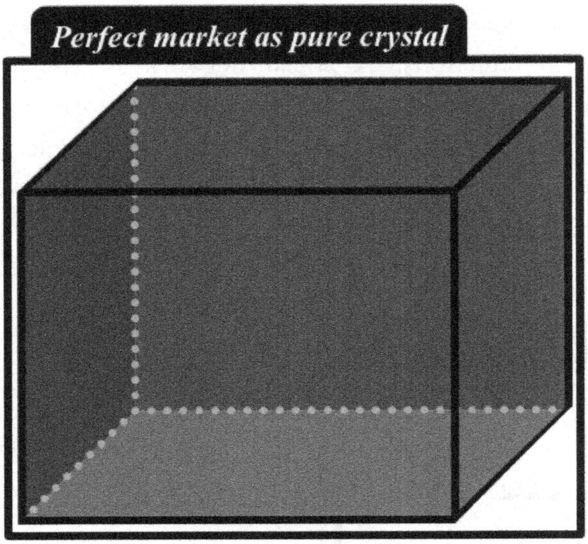


*Figure 9: The perfect market as a seamless, crystalline medium*

However, to scholars of economics, a more potent rendering is the negative image of Figure 8 (white on black): the market structure may be represented by the black square medium in Figure 9: everyone interacts with everyone, equally and neutrally.

This idyllic vision, however, is belied in real situations where pre-existing, waxing, and soon-to-emerge, as well as waning, vestigial social relationships shatter this crystalline structure (the perfect market). The resulting fractures enable information to flow faster, time cycles to shorten, etc., as illustrated in Figure 10.

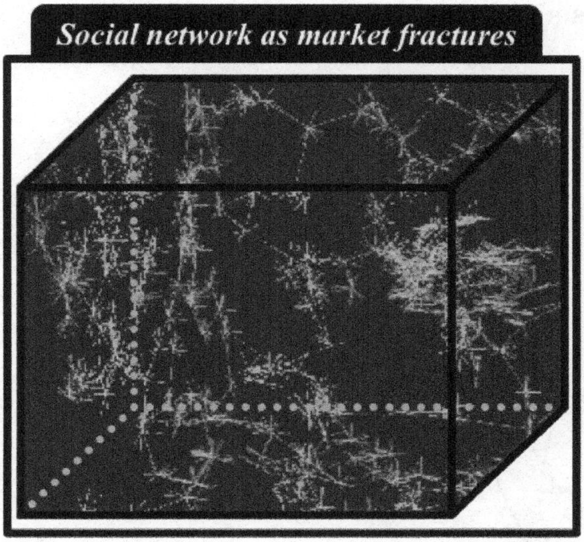


*Figure 10: A three-dimensional graphical rendering of a social network as fractures in an imperfect medium*

By shifting the analysis from the individual to the gene, Dawkins[94] demonstrates the wealth of insights a shift in level of analysis may reveal. Other illustrative examples come from fluid mechanics and optics. Respective insights into turbulence or interference have been gleaned by confronting the relative points of view of the fluid particle or the photon on the one hand or those of equipotent surfaces within the medium with the trajectories perceived by an objectively remote observer. Perhaps the most famous such shift in perspective is Copernican heliocentrism.

Organization studies often opt for individuals, teams, groups, whole organizations or industries as the level of analysis. This thesis proposes to use written language as a proxy for thought, as exercised within national cultures, large organizations and communities of practice.

---

[94] Dawkins, R. 1989a. The Replicators, *The Selfish Gene.* Oxford: Oxford University Press.

## "The safest general characterization of the European philosophical tradition is that it consists of a series of footnotes to Plato"[95]

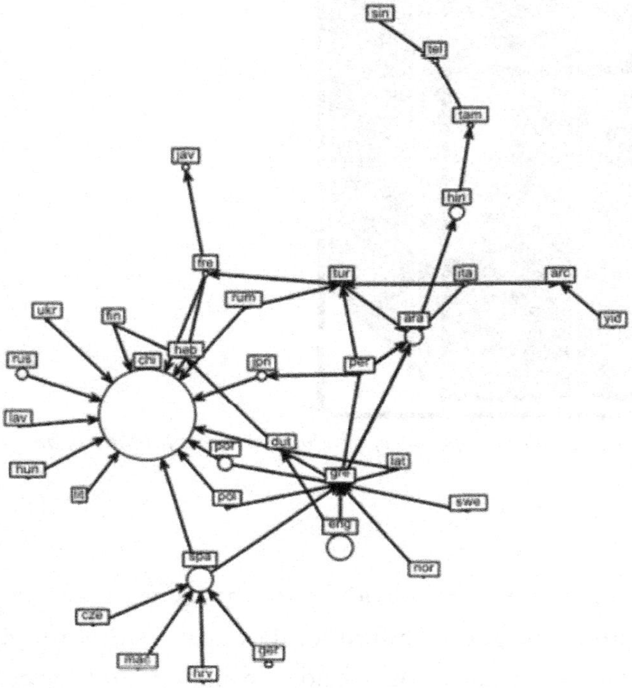


*Figure 11: A population-weighed extension to (Lieberman, 2009)'s directed graph of stereotypical incomprehensibility*[96]

---

[95] Whitehead, A. N. 1979. *Process and Reality*: Free Press.

[96] Figure 11 was produced based on primary data personally gathered to extend Lieberman, M. 2009. The directed graph of stereotypical incomprehensibility. (itself inspired by Rosenberg, A. 1979. The Hardest Natural Languages. *Lingvisticae Investigationes*, III(2): 323-339.), with nodes weighed according to worldwide speakers (Lewis, M. P. (Ed.). 2009. *Ethnologue: Languages of the World* (16th ed.). Dallas, TX: SIL International.) with R and its social network analysis package (Butts, C. T. 2008a. network: a Package for Managing Relational Data in R. *Journal of Statistical Software*, 24(2), Butts, C. T. 2008b. Social Network Analysis with sna. *Journal of Statistical Software*,

It can be understandably difficult to conduct tests separating Western thought and semantics in general when one's own cognitive thought processes are of Western tradition. Are P1-4 falsifiable[97] or circular arguments, given that they are formulated in Western thought and expressed in a Western language?

Many a philosopher has spent decades attempting to fathom whether the development of such abstract concepts as numbers, space, and time is contingent on human physiology: no attempt will be made here to further this line of inquiry.

François Jullien[98], a French philosopher whose initial training was in ancient Greek has spent his career attempting to contrast Western thought with the one surviving philosophical ancient tradition that has developed independently of it, i.e. Chinese thought.

As Figure 11 anecdotally illustrates, all but one culture faced with alien script attribute this writing to an alien culture, the culture they've encountered they feel most remote from their own. Written ciphers are mostly deemed Chinese (14 arrow ends) or Greek (10), to a lesser extent Spanish or Arabic. Whereas the Greek assume the unfathomable writing they encounter to be Chinese, the latter assume it to be otherworldly, i.e. "Heavenly script".[99]

---

24(6), Butts, C. T. 2009. sna: Tools for Social Network Analysis. R package version 2.0-1, Butts, C. T., Handcock, M. S., & Hunter, D. R. 2008. network: Classes for Relational Data. R package version 1.4-1. Irvine, CA, R Foundation for Statistical Computing. 2009. R: A language and environment for statistical computing. Vienna, Austria.) Node tags use the ISO-639 three-letter standard: International Standards Organisation. 2010. ISO 639-3 standard: SIL International.

97 Popper, K. 2002. *The Logic of Scientific Discovery* (15 ed.): Routledge.

98 Jullien, F. 1995. *Le détour et l'accès: stratégies du sens en Chine, en Grèce.* Paris: Grasset.

99 In Mandarin Chinese, incomprehensible writing is traditionally assumed to be Heavenly script (看起來像天書) whereas incomprehensible speech is traditionally assumed to be birdtongue (聽起來像鳥語) – neither are represented in Figure 11 as neither Heavenly script, nor birdtongue are catalogued as languages in Lewis, M. P.

**Chinese avoidance, Western confrontation.**

Jullien, in his Eastern forays, has contrasted Chinese avoidance with Western confrontation.

Though, like any idiomatic expression, the French « *C'est du chinois* ! »[100] should not be taken literally, according to Jullien it nevertheless marks Chinese thought as distinct. Be it strategy, poetry, painting, or philosophy, China traditionally cultivates the art of the lateral, the segue, the allusion. Chinese culture favors sideways approaches, whereas Greek culture – Western culture – engages front and center, favoring face-to-face confrontation.

Ancient Chinese military strategy is a case in point. Where the Greek martial model is the frontal contest between phalanxes, the Chinese seek victory without battle, by cunning and thwarting the plans of one's foe.

"Let others infer rather than explicitly state."[101]: Chinese diplomacy and rhetoric also take on indirect strategies. Jullien[102] illustrates his point by citing a series of negotiations that were exclusively conducted by citing classical poetry[103].

The pithy, seemingly shallow sayings of Confucius are only fully grasped as indicative clues: "The Master, in China, tells less than he indicates"[104]. In contrast to Socrates, for instance, Confucius crisscrosses his topics as if composing a fugue, never defining concepts such as 君子[105].

---

(Ed.). 2009. *Ethnologue: Languages of the World* (16th ed.). Dallas, TX: SIL International.

100 "It's Greek to me!"

101 «*Laisser entendre plutôt que dire.*»

102 Jullien (ibid.)

103 anon. 1879. *诗经 aka 詩經 (Shi Jing) aka The Book of Odes, Le Livre des Odes* (J. Legge, & M. Granet, Trans.).

104 « Le Maître, en Chine, parle moins qu'il n'indique. » Jullien (ibid.)

105 In Confucius. 2010. *论语 aka 論語 (Lun Yu) aka The Analects* (S. Couvreur, D. C. Lau, & J. Legge, Trans.)., Legge translates 君子 as "a man of complete virtue"; Lau as "gentleman"; Couvreur as « un homme honorable ».

In general, Chinese thought is immanent: it ignores the traditional Greek distinction between the concrete and the abstract, between the sensible and the intelligible; it is a philosophy of regulation, not of conversion; Chinese wisdom consists in joining the flow rather than warding off the cave's shadows[106] to see the light of truth.

**Western thought and semantics prime thinkers and speakers into attributing an a priori universal and absolute quality to enumerations.**

Joseph Needham[107] stresses China's independent philosophical development and contrasts it with its scientific development thus:

"There is, however, no doubt that China was in the circuit as far as the general diffusion of knowledge was concerned; indeed, there was probably far more exchange between China and its Western and southern neighbors than has often been supposed. Moreover, Chinese inventions and discoveries passed on in a continuous flood from East to West for twenty centuries before the scientific revolution, and the more technological they were the easier was their travel. The distinctive scientific ideas of the Chinese tended to stay at home, probably because their natural philosophy was not easily integrated in, or comprehensible to, the Westerners and their conceptions. Yet all these interchanges never affected the essential style of Chinese thought and culture patterns; which maintained a remarkable independence in spite of any diffusion. Contacts there were, interchanges of ideas and techniques certainly occurred, but never in such profusion as to affect the characteristic style of China's civilization, and hence of her science. In the light of this it is no exaggeration, then, to speak of the 'isolation' of China, or, alternatively, of the rest of the world from her; culturally that was something very real."

---

106 Plato. 1999. *Great Dialogues of Plato: Complete Texts of the Republic, Apology, Crito Phaido, Ion, and Meno,* Vol. 1 (Warmington and Rouse ed.). New York: Signet Classics.

107 Ronan, C., & Needham, J. 1978. *The Shorter Science and Civilization in China: An Abridgement of Needham's Original Text.* Cambridge: Cambridge Univerity Press.

I shall test P1 by exploring ancient Chinese mathematics notably the following works:

- The Suànshùshu[108], the earliest known extensive Chinese treatise on mathematics, is an ancient Chinese collection of writings on mathematics approximately seven thousand characters in length, written on 190 bamboo strips. It was discovered together with other writings in 1983 when archaeologists opened a tomb at Zhangjiashan, Hubei province. From documentary evidence this tomb is thought to have been closed in 186 BCE, early in the Western Han dynasty. The occupant of the tomb —whose name is unknown to us— appears to have been a minor local government official, who had begun his career in the service of the Qin dynasty, but started work for the Han in 202 BCE. The Suanshushu is anonymous, in the sense that we do not know the name of the person who assembled this material.
- The I-Ching[109], one of the most important books in the Chinese literature has origins in mythical Antiquity. Nearly all that is greatest and most significant in the three thousand years of Chinese cultural history has either been inspired by this book, or exerted an influence on the interpretation of its text, including the two main branches of Chinese philosophy, Confucianism and Taoism. It is included in this dissertation because its foundation is binary.

---

108 Cullen, C. 2004. *The Suàn shù shu 筭數書 'Writings on reckoning': a translation of a Chinese mathematical collection of the second century BC, with explanatory commentary*. Cambridge: Needham Research Institute.

109 anon. 1967. *易经 aka 易經 (Yi Jing), aka I Ching, the Book of Changes* (R. Wilhelm, C. F. Baynes, & E. Perrot, Trans.).

- The works of the Logicians or School of Names (名家), a Chinese philosophical school that grew out of Mohism in the Warring States Period and went extinct during the subsequent Legalist Qin Dynasty. Historians attribute this to two factors: because the political and intellectual climate of the new empire was hostile to their purely theoretical, occasionally flippant inquiries, and partly because with unification their political services became obsolete. How Deng Xi, Yin Wen, Hui Shi (惠施 aka 惠子), Gongsun Long (公孙龙), Cheng-gong Sheng, Huang Gong, and Mao Gong and perhaps Huan Tuan and others[110] grasped numbers will be investigated.

Secondary sources will also be studied, including:

- Chinese publications in English explaining Ancient Chinese logic and mathematics[111];
- Chinese scholarly curricula in Chinese of Western mathematical and logical thought;
- Western studies of Chinese thought[112].

Finally, Chinese scholars were interviewed: Professors Liu Fang and Wu Bolong of the Chinese Academy of Social Sciences; Professor Song Jijie of Tsinghua University; Professor Liu Chang of Peking University. Some insights were later crosschecked with Professor François Jullien of Université Paris Diderot, Professor Paul Jorion and Alain Butler.

---

[110] this list of Chinese thinkers was drafted after perusing Fraser, C. 2009. School of Names. In E. N. Zalta (Ed.), *The Stanford Encyclopedia of Philosophy*, Summer 2009 ed.

[111] e.g. Xiao, J., & Li, J. 2008. *An Outline History of Chinese Philosophy* (S. Zhang, Trans.). Beijing: Foreign Languages Press.

[112] e.g. Granet, M. 1934. *La pensée chinoise.* Paris: Albin Michel.

**Western thought and semantics prime thinkers and speakers into attributing an a priori universal and relative quality to spatial information.**

P2 will be tested likewise, by exploring the same ancient Chinese works:

- The Suànshùshu.
- The I-Ching.
- The reported works and thoughts of members of the School of Names. How the Chinese Logicians grasped space will be investigated.

Secondary sources studied for P1 will also be studied for P2. The Chinese scholars queried for P1 will also be queried for P2.

**Western thought and semantics prime thinkers and speakers into attributing an a priori universal and relative quality to temporal information.**

The same ancient Chinese works will be explored in testing P3:

- The Suànshùshu.
- The I-Ching.
- The reported works and thoughts of members of the School of Names. How Chinese Logicians grasped time will be investigated.

Secondary sources studied for P1 and P2 will also be studied for P3. The Chinese scholars queried for P1 will also be queried for P3.

**Western thought and semantics prime thinkers and speakers into attributing a subjective, shared, and social quality to coded information.**

P4 is already well established as a proposition[113] and is merely put forth here for the sake of completeness.

[113] Berger, P. L., & Luckmann, T. 1966b. *The Social Construction of Reality: A Treatise in the Sociology of Knowledge*: Anchor Books.

# IV] RESULTS —UNFUNGIBLE NUMBERING, SPACETIME

### A word on fungibility

Fungibility is a concept, borrowed from financial services, and introduced here. Two objects are deemed fungible to a third if the latter views the former as indistinguishable members of the same set. E.g. equally denominated banknotes are fungible to petty cash managers, but not to their issuer (the former will ignore their serial numbers; the latter will not); if a company provides staff with a pool of durable resources (cars, chairs), the staff using them will mostly see them as fungible and the staff managing them should be tracking them individually (how else would you know which car to service and when, or that it's always the same chair that breaks a leg?).

Fungibility, as a concept, is pervasive to management. In fact, viewing two or more objects as fungible leads to their commensuration[114].

### The Ancient and Modern Chinese use several non-fungible numbering systems

Examination of the I-Ching, (Granet, 1934)[115], corroborated by the interviews of Prof. Liu and Wu reveal the following —striking to this Western author— results.

---

114 Espeland, W. N., & Stevens, M. L. 1998. Commensuration as a Social Process. *Annual Review of Sociology*, 24: 313-343.

115 Granet, M. 1934. *La pensée chinoise*. Paris: Albin Michel.

The ancient Chinese used 3 different sets of non-fungible numbers —in different bases!— depending on the context:

- base 10 *numbers of the Heavens*: 甲, 乙, 丙, 丁, 戊, 己, 庚, 辛, 壬, 癸;
- base 12 *numbers of the Earth*: 子, 丑, 寅, 卯, 辰, 巳, 午, 未, 申, 酉, 戌, 亥;

*base 2 or 64* divination numbers *in the Yi Jing (see Figure 12, which brings ring theory to mind).*

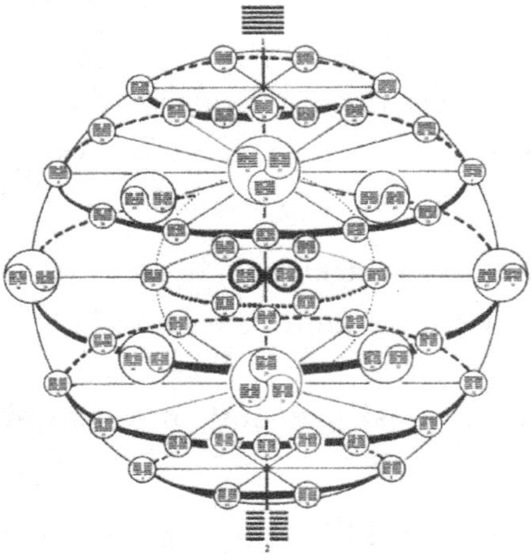

*Figure 12: A state-transition graph with the 64 hexagrams*[116]

Indeed, Sino-Western exchanges led to the introduction before the Christian era of a fourth numbering system for West-inspired numbering, also base 10: 〇, 一, 二, 三, 四, 五, 六, 七, 八, 九, 十 whose writing was resp. complicated into 零, 壹, 贰, ▪ or 参, 肆, 伍, 陆, 柒, 捌, 玖, 拾 credited to China's only female ruler, Empress Wu Zetian (7th c. AD), to prevent public accounting fraud.

[116] anon. 1967. *易经 aka 易經 (Yi Jing), aka I Ching, the Book of Changes* (R. Wilhelm, C. F. Baynes, & E. Perrot, Trans.).

These numbering systems are used to this day.

- *Numbers of the Heavens* are ordinal; e.g. the H5N1 virus is H戊N甲, hepatitis B is 乙肝.
- *Numbers of the Earth* are used in cyclical contexts; e.g. 申 represents 240º (approx. Southwest) to contemporary Chinese navigators, and has done so at least since the 14th century; Chinese-speaking pilots would use 申 similarly to English-speakers' "I have bogeys at my 8 o'clock"; more vestigially, numbers of the Earth subdivide the day into twelve 2-hour periods starting with 11pm-1am.
- Phone numbers, most measures use the 4th, Western inspired, numbering system, e.g. 八六一〇六三〇七〇八五〇; Arabic numbering is used equivalently, e.g. 12024561414.

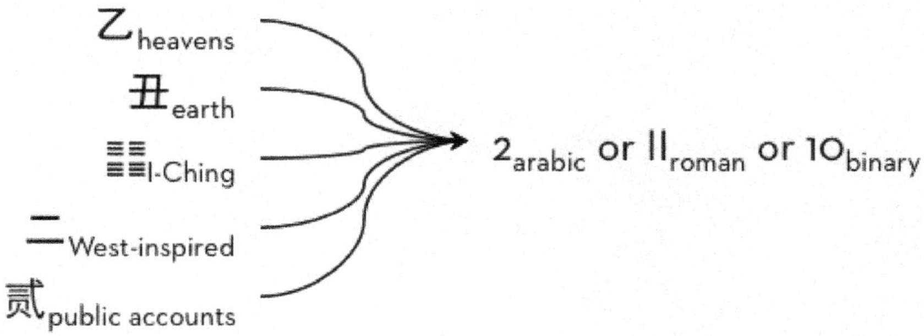


*Figure 13: Translation of Chinese numbers to Western numbers is non-injective*

As Figure 13 illustrates, translation of Chinese numbers to Western numbers is therefore non-injective.

**The Ancient Chinese had a single, unified (to Westerners) notion of space-time**

The ancient Chinese did not distinguish spatial and temporal information, but exercised spatiotemporal holism[117]. Timing applied within a limited geography. Changing kingdoms implied a change of calendar; even changing kings reset calendars and maps … after a proper mourning period. This is illustrated in HuiShi's paradoxical "今天我去了越国□然后昨天我从越国回来了"[118].

Translations of (a) Western spatial information to Chinese spatio-temporal information and of (b) Western temporal information to Chinese spatiotemporal information are therefore non-injective.

---

117 Granet (ibid.)

118 "I set off for Yueh today and came from there yesterday" HuiShi (Watson, tr.).

## Possible objections

### On method

Historical data analysis is biased.

> All methods are biased. In too many instances to be dismissed, newly uncovered historical documents contradict previous historic consensus. That said, the author cannot conceive of any historical document that would reverse one of the conclusions presented here. Of course, this may merely be a symptom of the author's lack of imagination.

Arguments on Western thought, expressed in English are either tautological or circular.

> This is a fair comment. There are limits to self-reflection. Operating simultaneously on several levels is a delicate exercise. Semioticians have the luxury of being able to use quotes to distinguish the word 'word' from its meaning, i.e. word —without quotes.

Ancient Greek texts no longer exist. Ancient Chinese texts have in some cases been preserved. Any comparison is moot.

> Indeed, when Aristotle or Plato are quoted, it no longer is from the original texts, but from translations, i.e. interpretations, themselves translated again over a thousand years after their deaths (e.g. from Hebrew or Arabic into, say, Latin). If there is anything this thesis cautions against, it is thinking one's frame of reference is universal. Over-interpretation of translations of translations of translations of Greek texts is a potential pitfall here.

Conversely, but not unsimilarly, some Ancient Chinese texts have survived two millennia. Their interpretation, as well as second hand commentaries of lost texts, still fuels scholarly debate. Chinese characters have evolved over time and leave room for varied interpretations. Texts, such as the Analects, hint obliquely rather than inform in a documentary fashion.

In both contexts, one cannot rule out that some documents may have been willingly destroyed or stealthily censored by displeased authorities, given the power over the flow of information the Catholic Church and Legalist China exercised in the late first and early second Millennia in their respective spheres of influence.

The concepts of numbers, space and time do however seem more immediately amenable to comparative scrutiny than more immaterial concepts such as honor, freedom, debt, or self, for which ignorance of the context in which they were uttered is rather more difficult to project in contemporary Europe or in contemporary China.

Besides, P1-4 are valid in English, French and, as far this author knows, all living Indo-European languages. Confronting them with contemporary Chinese remains relevant even in the absence of Aristotle's original texts.

## On conclusions drawn

Having different numbering systems does not mean having different numbers: 42 = XLII.

> This, too, is a fair point, which mirrors Dehornoy's musing on Martian math. However, it does not invalidate the argument that having a single numbering power of expression may induce one to compare incomparable measures, whereas having a plurality of numbering systems to segregate cognitively the arithmetic, the ordinal, the cyclical from each other and the phenomenological from the socially constructed may hinder uncalled-for amalgamation.

The uncovered numbering systems may be vestigial traces of proto-Sino cultures, e.g. Southern Song dynasty Suzhou numerals (〇, 〡 or 一, 〢 or 二, 〣 or 三, 〤, 〥, 〦, 〧, 〨, 〩, 拾) are still in use in markets, for instance; why could the three ancient numbering systems not have been juxtaposed when different proto-Chinese cultures merged?

> In view of the contexts in which the various numbering systems are used, the author believes this to be unlikely. Clearly, ordinal numbers could be used in cyclical contexts (angles, time subdivisions…) and binary systems can be applied to any coding as computing attests. Nevertheless, future research could delve further into Chinese Antiquity.

Having different spatial systems does not mean spatial experience is not universal: up, down, east, west, north, south still bring structure; having different timing systems does not mean temporal experience is not universal: days, months, years still bring structure.

> The author does not wish to set up a budding civilization in a gravity-less space pod or sun-less cavern to see what happens. The insight presented does not imply space or time orientation, organization, or subdivisions are universal. The Chinese, lacking the benefit of framed Platonic thought, idealizing objects, saw time as a superposition of cycles, finite in space, whereas Westerners conceptualized the infinite, geometric straight line —a dubious benefit, at that, since one can wonder whether the discovery of relativity might not have been hindered by the semiotic decoupling of space and time in Western thought.

Aren't 甲, 乙, 丙, … merely equivalent to first, second, third, and so on in English? And aren't 子, 丑, 寅, … similarly equivalent to twelfths of periods or full angles?

> Just as a verb and a substantive (i.e. to operate and an operation) may share a common meaning, in English, first and one, as well as prime and a, may be different words but have a common symbolic meaning. Semiotically they are the same. Casting one word or another in a sentence will be contingent on the role the underlying semiotic concept has to play within the sentence. In Chinese, however, 甲, 一 or 子 can't be cast in sentences depending on their role in the sentence in question. Besides, the fact they are in different bases reinforces their non-substitutability.

## On lack of relevance to contemporary managerial issues

According to Prof. Song, contemporary Chinese thought has been strongly influenced by Western thought since at least 300 years. How does studying texts a few scholars study help advance management science?

> Beyond the implications discussed below, and despite the relative Americanizing of the Chinese manager, some of the texts perused are acknowledged as central to Chinese culture, most notably the Yi Jing and Shi Jing.

## Discussion

I intended to prove: (a) numbers, space and time are social constructs; (b) philological methods shed light unmatched by past multicultural surveys.

Surveys, like statistics, inform of the surveyors more than of the survey object. Lexical gaps between cultures or professions thus revealed may be symptoms of symbolic gaps, especially in the case of mathematical objects (from algebra, geometry, calculus…) which have no phenomenological grounding; lexical gaps revealed may also be symptoms of semiotic gaps, or polysemy. Examples are given below.

The managerial and organizational implications inferred thus are:

> All information, even quantified, needs to be contextualized. Both the conventions selected (to measure and code, including count) and the institution endorsing the convention should qualify each shred of information.
>
> The more self-doubt and open-mindedness exercised in a cross-cultural or cross-disciplinary context, the greater the awareness of the context.

In cross-cultural, cross-disciplinary settings, irrespective of the level (individual, team, organization, industry or profession), exploring whether what is known is still valid in uncharted land, and not merely exploiting it is a key organizational, managerial, or personal skill.

Understanding your counterparts means learning to unlearn what you did not know is no longer valid!

# V] IMPLICATIONS FOR QUANTS — WHAT YOU THINK YOU THINK

**iii the case of the trusted engineer**

When automakers Ginger and Lime merged their operations, the newly merged Linger's executive committee tasked Alex, Lime's chief engineer, to propose a common transmission system for Linger's next generation of mid-market vehicles.

Nine months later, Alex submits a 250-page report that is circulated to the committee members. During their next meeting, Lee, the GM, brings up the point: "Point number 5: our new midmarket transmission platform. Has everyone read Alex's report?" Lee pauses, looks around. "Right, Alex, go ahead."

Alex gets up and starts to speak "I would first like to thank all the people who contribu..."

Lee interrupts "No, Alex, you don't understand. Your proposals are approved. You can go ahead. Right. Point number 6: ..."

Recent events have made it clearer than ever that monetary, economic, financial, and business models have had an impact on society as a whole, an impact not necessarily deemed positive. Models of social phenomena seem to fail just when we need them the most.

Is this an in-built flaw in the modeling process? Should we drop our heavier tools as Weick[119] suggests in order to survive the forest fire that has engulfed organization theorists? Or is the devil in the details, i.e. are the mathematical building blocks modeling relies on where the flaws lie? Are we, as Maslow[120] suggests, "*treating everything as if it were a nail because the only tool we have is a hammer*"? Or is our blind trust in numbers[121] at the core of our present shortcomings?

Crosby[122]describes "*late Medieval*[123] *Westerners* [as] *monotheists fascinated with light,* [who] *gloried in pantometry*[124]" and exposes how quantification changed Western civilization. Jorion[125] furthers this analysis by anthropologically tracing the birth of *Truth* to Ancient Greece, as evidenced by the Principle of Contradiction[126], and the birth of *(objective) Reality* to the 16th and 17th centuries, Reality whose systematic, quasi-viral replication produced, of course, modern Positivism and led to the parochial Postmodernism reaction against it in the West[127].

---

119 Weick, K. E. 1999. Theory construction as disciplined reflexivity: tradeoffs in the 90s. *Academy of Management Review*, 24(4): 797-806.

120 Maslow, A. H. 1966. *The Psychology of Science: A Reconnaissance*. New York: Harper & Row.

121 Porter, T. M. 1995. *Trust in Numbers*. Princeton: Princeton University Press.

122 Crosby, A. W. 1997. *The Measure of Reality: Quantification and Western Society, 1250-1600*: Cambridge University Press.

123 Naming periods such as "the Middle Ages", "the Renaissance"... is a matter of debate among historians, a debate within which this author has no intention of taking sides. Admittedly, a numbered year such as "1492 BC" or "52 CE" is as much a social code as "the Iron Age", "the Enlightenment". However, there is broad consensus and no debate on the former two, as it merely is a spatially bounded timing code. So Late Medieval here means of the period studied by Crosby, i.e. 1250-1600.

124 Pantometry, a term coined by Crosby, means universal measurement.

125 Jorion, P. 2009. *Comment la vérité et la réalité furent inventées*. Paris: Gallimard.

126 Aristotle. 2000b. *Métaphysique* (J. Tricot, Trans.). Paris: Vrin.

127 Adler, N. J. 1983. A Typology of Management Studies Involving Culture. *Journal of International Business Studies*, 14(2): 29-47.

This pantometric *mentalité* pervading contemporary thought to this day is epitomized in the learning and education community by Thorndike[128]'s oft-quoted "*whatever exists at all exists in some amount. To know it thoroughly involves knowing its quantity as well as its quality*" and, closer to our field, Kuhn[129]'s "*numbers register the departure from theory with an authority and finesse that no qualitative technique can duplicate, and that departure is often enough to start a search*". Mangin[130], in his breakdown of performance management systems design, aptly observes "*decisions are about premises, since logical and mathematical deduction leaves no room for human discretion: once the problem is (politically) well defined, it is (logically) already solved*".

Jorion[131], Jullien[132], and I hereabove, with distinctive methods, reveal that Chinese and Western thought developed in Antiquity independently and along different paths, including in such fields, as arithmetic and geometry, the universality of which Westerners tend to take for granted. Jullien[133] notably establishes "*Chinese mathematics always are* applied *mathematics*" and "*the very notion of* ideal —and the process of idealization, i.e. modeling— *is idiosyncratic to Europe*".

As previously mentioned, behavioral observation of social science modelers reveals:

---

128 Thorndike, E. L. 1918. The nature, purposes, and general methods of measurements of educational products. In G. M. Whipple (Ed.), *The Seventeenth yearbook of the National Society for Study of Education. Part II. The Measurement of Educational Products*: 16. Bloomington, IL: Public School Publishing Co.

129 Kuhn, T. S. 1961. The Function of Measurement in Modern Physical Science. *Isis*, 52(2): 161-193.

130 Mangin, N. 2008. A comprehensive analytical framework for studying performance evaluation, *2008 Academy of Management Annual Meeting*. Anaheim, CA: Academy of Management.

131 Jorion, P. 2009. *Comment la vérité et la réalité furent inventées*. Paris: Gallimard.

132 Jullien, F. 2009. *L'invention de l'idéal et le destin de l'Europe*. Paris: Seuil.

133 Jullien (ibid.)Jullien, F. 2009. *L'invention de l'idéal et le destin de l'Europe*. Paris: Seuil.

(a) mathematics are a social construction[134];

(b) practitioners close to the core of axioms involved in a particular theory understand the assumptions made better than the many more people applying mathematics;

(c) application of mathematics conforms to the *garbage can model*[135] as observation of phenomena and availability of the mathematical model occur in a concurrent time frame;

(d) models are adopted to make sense of phenomena[136], not so much because of their relevance …

…but rather because (e) applying mathematics in an organizational context may lead to their mimetic application in peer organizations[137];

(f) adoption by trendsetters then propagates consistently with fashion theories[138];

---

134 Berger, P. L., & Luckmann, T. 1966b. *The Social Construction of Reality: A Treatise in the Sociology of Knowledge*: Anchor Books, Piaget, J., & Szeminska, A. 1941. *La genèse du nombre chez l'enfant.* Neuchâtel: Delachaux & Niestlé.

135 Cohen, M. D., March, J. G., & Olsen, J. P. 1972. A Garbage Can Model of Organizational Choice. *Administrative Science Quarterly*, 17(1): 1-25.

136 Weick, K. E., & Bougon, M. 1986. Organizations as cognitive maps: charting ways to success and failures. In H. P. Sims, Jr., & D. A. Gioia (Eds.), *The thinking organization*: 102-135. San Francisco: Jossey-Bass.

137 DiMaggio, P. J., & Powell, W. W. 1983. The Iron Cage Revisited: Institutional Isomorphism and Collective Rationality in Organizational Fields. *American Sociological Review*, 48(2): 147-160.

138 Abrahamson, E. 1996. Management fashion, academic fashion, and enduring truths. *Academy of Management Review*, 21(3): 616, Kieser, A. 1997. Rhetoric and Myth in Management Fashion. *Organization*, 4(1): 49, Trepo, G. 1987. Introduction and diffusion of management tools: the example of quality circles and total quality control. *European Management Journal*, 5(4): 287.

(g) widespread adopters, chronologically, geographically, or topologically distant from the community that set the underlying mathematical concepts have a weaker understanding of the axiomatic underpinnings of these concepts and approach their application with the machine-like regulation of bureaucracy[139];

(h) mathematical concepts will be legitimized by repeated enactment, and social change may both emerge from and influence their legitimation[140];

(i) groups which succeed in imposing their discourse shape peoples' understanding and gain a powerful advantage over groups with which they are in normative conflict[141].

---

139 Weber, M. 1991. *Essays in Sociology* (1st ed.): Routledge.

140 Granovetter, M. 2007. The Social Construction of Corruption. In V. Nee, & R. Swedberg (Eds.), *On Capitalism*: 152-172: Stanford University Press.

141 Ewick, P., & Silbey, S. S. 1995. Subversive Stories and Hegemonic Tales: Toward a Sociology of Narrative. *Law & Society Review*, 29(2): 197.

## Safety in Numbers ?

## What lies beneath ?

| | Situational variables | Data properties | Methodology issue | Error source | Error | Managerial state |
|---|---|---|---|---|---|---|
| 1 | Few products or service transactions | Small number of observations | | Sampling error | Random variation | Exaggerated variability, causal complexity |
| 2 | Divisionaliz-ation | Data dis-aggregation | Small number of observations | Sampling error | Random variation | Exaggerated variability, causal complexity |
| 3 | Small size of country | Small size of organization | Small number of observations | Sampling error | Random variation | Exaggerated variability, causal complexity |
| 4 | Multiple causes | Small correlation | | Sampling error | Random variation | Exaggerated variability, causal complexity |
| 5 | | All variables measured with error | | Measurement error | Understated correlation | Underuse of cause |
| 6 | | | Few items | Measurement error | Understated correlation | Underuse of cause |
| 7 | High positive correlation between sales and costs | Profit as the variable | Difference score | Measurement error | Understated correlation | Underuse of cause |
| 8 | Isomorphism of survival | Missing cases | Selection | Range restriction | Understated correlation | Underuse of cause |

| Situational variables | Data properties | Methodology issue | Error source | Error | Managerial state |
|---|---|---|---|---|---|
| 9 | Extreme cases | | Range extension | Overstated correlation | Overreliance on cause |
| 10 Multiple causes | Small true correlation | Confounding more feasible | Confounding | Overstated / understated correlation | Overreliance / underuse of cause |
| 11 | Correlation between focal cause and other factor | Masking or inflating | Confounding | Over / understated correlation | Overreliance / underuse of cause |
| 12 Correlations of sales and costs with profits | Profit | Definitional connections | Confounding | Over / understated correlation | Overreliance / underuse of cause |
| 13 Performance-driven organizational change | Negative correlation of cause and performance | Reverse causality | Confounding | Making of true positive correlation | Mistaken causality |
| 14 Large organizations | Organizational experiments | | Bias | Understates effects | Self-fulfilled skepticism |

*Table 2: Summary of Statistico-Organizational Theory*[142]

[142] Donaldson, L. 2010. *The Meta-Analytic Organization, introducing Statistico-Organizational Theory*. Armonk, NY: M.E Sharpe.

Debate on mathematics in social science is often focused on the proper application of mathematics, most notably on rigorously applying statistics and probabilities. Donaldson[143] suggests the one principle management scholars agree upon —and even share with practitioners— is the concept that statistical, numerical measurements are useful for understanding organizations. Donaldson further exposes how the current numerical data collection methods employed by both management academia and practitioners present structural flaws.

Sometimes, new mathematics tools are introduced in an effort to describe phenomena previously less successfully modeled —e.g. Erdős[144] in network analysis, or Mandelbrot[145] in stock valuation.

But rarely are the core tenets of mathematics questioned. Behaviorally speaking, this is unsurprising. As cognitive distance from a theory increases, so does the misunderstanding of its field of validity. For instance, a macro-economist is more likely to question a macro-economic theory than a mathematical axiom he nonetheless needs to support his theory. Indeed, the further one is cognitively from a theory, the more likely one is to apply it dogmatically, and outside its domain of validity.

However, questioning core mathematical tenets is far from fruitless. Since Euclid, geometry is based on five postulates, illustrated in Table 3. Euclid's fifth postulate was long considered weaker than the first four and many a mathematician attempted to prove the fifth postulate from the first four. Some of them even succeeded with proofs that were later demonstrated to be methodologically flawed[146].

---

143 Donaldson, L. 2010. *The Meta-Analytic Organization, introducing Statistico-Organizational Theory*. Armonk, NY: M.E Sharpe.

144 Erdős, P., & Rényi, A. 1959. On Random Graphs. *Publicationes Mathematicae Debrecen*, 6: 290-297.

145 Mandelbrot, B. B., & Hudson, R. L. 2004. *The (mis)behavior of markets: a fractal view of risk, ruin, and reward.* New York: Published by Basic Books.

146 Bourguignon, J.-P. 2010. Espaces courbes de Gauss à Perelman, en passant par

| **Euclid's postulates[147]** | |
|---|---|
| 1. | A line may be drawn between any two points. |
| 2. | Any line segment may be extended indefinitely. |
| 3. | A circle may be drawn with any given point as center and any given radius. |
| 4. | All right angles are congruent. |
| 5. | If two lines are cut by a transversal in such way that the sum of the two interior angles on one side of the transversal is less than two right angles, then the lines meet on that side of the transversal. |
| **Playfair's postulate aka the parallel postulate[148]** | |
| | Through a point not on a given line, exactly one line can be drawn in the plane parallel to the given line. |

*Table 3: The five Axioms of Euclidian geometry and Playfair's Postulate*[149]
*N.B.: Euclid's fifth postulate is equivalent to Playfair's postulate.*

We now know the fifth postulate to be pivotal. If it is true, we have classical Euclidean geometry. If it is false, we enter the realm of non-Euclidean geometry, which theoretical physicists have found useful in relativity. More prosaically, software engineers have also applied it to index rich media (pictures and movies). Non-Euclidean geometry divides into (a) hyperbolic spaces where several lines can be drawn through any point not on a given line parallel to the given line in a plane in some cases; and into (b) spherical spaces where parallel lines can share two or more points without being equal.

---

Einstein. In S. M. d. France (Ed.), *Un texte, un mathématicien*.

[147] Euclid. 1956. *The thirteen books of the elements* (T. L. Heath, Trans.): Dover Publications.

[148] Playfair, J., & Euclid. 1860. *Elements of geometry*: J.B. Lippincott & Co.

[149] Noronha, M. H. 2002. *Euclidean and Non-Euclidean Geometries*: Prentice Hall.

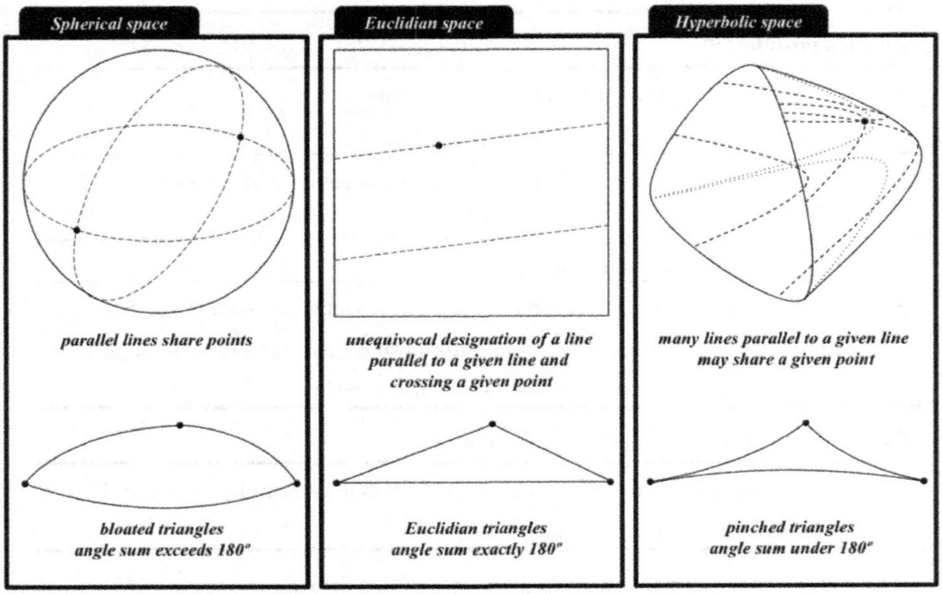


*Figure 14: Illustration of Spherical, Euclidian and Hyperbolic spaces*

Social scientists use geometry sparingly. Rare examples include the development of cultural or pseudo-spatial distances[150] that, apart from their statistical foundations, rely on distances.

---

150 Hofstede, G. 1983. National Cultures in Four Dimensions: A Research-Based Theory of Cultural Differences Among Nations. *International Studies of Management and Organization*, 13(1,2): 46.
Puffert, D. J. 2002. Path Dependence in Spatial Networks: The Standardization of Railway Track Gauge. *Explorations in Economic History*, 39(3): 282-314.

Numbers, however, are used daily.

| | |
|---|---|
| **Peano's Postulates** Let *N* be a set and *S* a relation on $N \times N$. If $n = S(m)$, then *n* is said to be a successor of *m*. If *N* satisfies the following axioms, *N* is called the set of *positive integers*. | |
| **Axiom 1** | The relation *S* is a function; that is, if *m* belongs to *N*, then there exists a unique *n* belongs to *N* such that $n = S(m)$. |
| **Axiom 2** | There exists an element 1 in *N* such that for all*n*s belonging to *N*, $1 \neq S(n)$; that is, 1 is not in the range of *S*. |
| **Axiom 3** | *S* is one to one; that is, if $n = S(m)$ and $n = S(m')$, then $m = m'$. |
| **Axiom 4** | If *M* is a subset of *N* which has the following properties: (a) 1 belongs to *M* (b) if *m* belongs to *M* and $n = S(m)$, then *n* belongs to *M*, then $M = N$. |

*Table 4: The four Axioms of Number Theory*[151]

Just as Euclidean and non-Euclidean geometry rests on five axioms, number theory —sometimes called Peano arithmetic— rests on four, illustrated in Table 4. In an interesting parallel, the last axiom differs in nature from the others. The fourth axiom of number theory —i.e. the induction axiom, or the Dedekind–Peano axiom— states:

> If 'statement S is true for a base case' —e.g. 0 or 1— and induction is shown, —i.e. "for all ns, 'S true for some number n' implies 'S true for n+1'"— then S is true for all numbers[152].

[151] Anderson, J. A., & Bell, J. M. 1997. *Number Theory With Applications*: Prentice Hall.

[152] Dedekind, R. 1888. *Was sind und was sollen die Zahlen? (What are and what should the numbers be?)*: Braunschweig.

The first three are of first-order logic, whereas the induction axiom is of second-order logic; i.e. it relies on semantic generalizations such as "*and so on*" whereas the first three do not. The mirror image —in the sense that non-Euclidean geometry mirrors Euclidean geometry— to Peano arithmetic is Robinson arithmetic[153]. In Robinson arithmetic, proving (building addition from the increment function, and multiplication from addition) *6 times 7 equals to 7 times 6* is possible; however, proving the more general *x times equals y times x for all xs and all ys* is not. In other words, proving a statement for *any* number incrementally reachable from 0 does not make that statement S true for *all* numbers.

153 Robinson, R. M. 1950. *An Essentially Undecidable Axiom System.* Paper presented at the International Congress of Mathematics, Cambridge 1950.

# VI] NEW PATHS FOR SOCIAL MODELING —RE-THINKING THINKING

**iv the case of the bullion-dragging device**[154]

Long ago and far away, there was a wealthy land that invested a large portion of its resources in building huge conveyances to facilitate bullion-based trade. Because the wheel had not yet been invented, teams of slaves were needed to drag the devices around. The Federal Research Bureau of the Transportation Department formed a council to find a device that would make dragging easier. The council tested and discarded a series of designs.

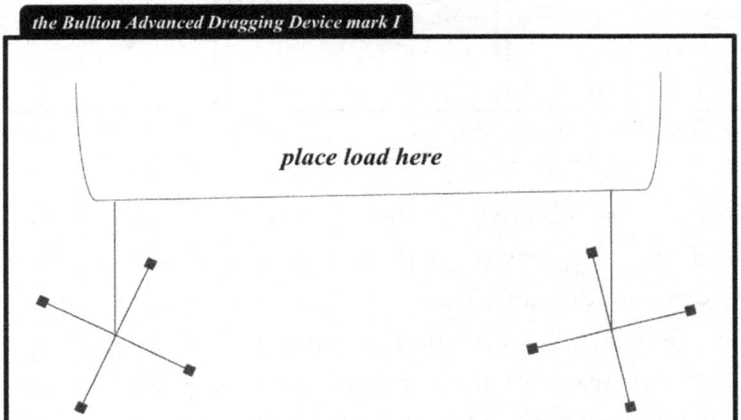


*Figure 15: The first Bullion Advanced Dragging Device*

The first was a simple long cross with an axle (cf. Figure 15). The concept was to reduce the energy lost to friction. The device was to rotate as the load was dragged.

[154] This is heavily inspired by Pamas, D. L. 1984. is ADA Too Big? *Communications of the ACM*, 27(11): 1155.

This device was a success for lighter loads but could not stand the weight of the land's bullion-based currency of choice. An improved model was therefore necessary.
This device could handle heavy loads but a great deal of effort was required to lift the device in order to rotate it. Four intermediate supports were introduced (cf. Figure 16b).
Unfortunately, these supports occasionally cracked and required further reinforcement. Every joint was encased in a metal patch and iron straps were nailed in each support (cf. Figure 16c).

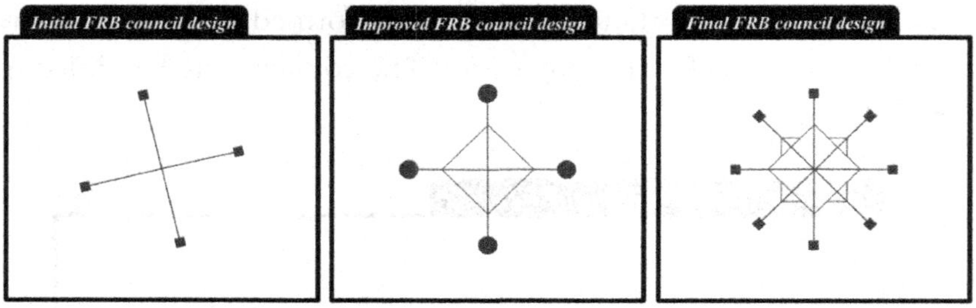


*Figure 16: Successive FRB council designs*

The design was accepted by the FRB and the TD as the Bullion Advanced Dragging Device (BADD) and promulgated as a standard. It quickly came under attack by engineers and traders for being too complex, too heavy, too difficult to manufacture, and too difficult to use. One of the members of the council[155] wrote a paper showing beyond doubt that if any piece of wood were removed the structure would collapse.

[155] In their respective fields: Mandelbrot, B. B., & Hudson, R. L. 2004. *The (mis)behavior of markets: a fractal view of risk, ruin, and reward.* New York: Published by Basic Books. Weick, K. E. 1999. Theory construction as disciplined reflexivity: tradeoffs in the 90s. *Academy of Management Review*, 24(4): 797-806.

The wheel would not be invented for another thousand years. During that millennium, many were kept fully employed working on BADD-related activities (research, manufacturing, rating agencies, maintenance…). Each time they added a supporting arm, they were able to demonstrate improvement but also to show that further research and further, more refined field monitoring was needed.

## Reflections on cognitive processes

### On the Social Nature of Information

No information or identity regarding any individual exists outside a social context.

Even one's first name, surname or age exist only within predetermined conventions and only if socially sanctioned authorities enact these conventions and institutionally acknowledge in official registries natural events such as births. E.g. in France, these conventions include the attribution of a first name, the transmission of a surname through a filial connection, and the coding of a relative interval between two dates according to a pre-agreed calendar.

Conversely, as nodes in social networks, all individuals are vehicles for shared languages, information, and contexts. In a sense, all individuals are a medium.

### On Impression Management

Impression management[156] puts forward the notion that all actions an individual undertakes in society are less expressions directed towards others than they are attempts to control the impressions one make on others; indeed, the only peers one doesn't try to impress are people one considers non-persons.

[156] Goffman, E. 1959. *The Presentation of Self in Everyday Life*: Anchor Books.

Goffman notes in his seminal thesis: "*In social intercourse, one (a) presents oneself and one's activities to others, attempts to guide and control the impressions others form of one, (b) employs certain techniques in order to sustain one's performances, (c) is driven to manage one's impressions by fear ranging from slight embarrassment to deep humiliation.*"

For Goffman, impression management rules actual expressions. Whether one expresses *real* attitudes, beliefs, or emotions, information conveyed by language, conduct, or appearance will be impressed upon one's audience, be it another individual, a group, or society at large, depending on the context of the performance. Rightly so, the impression conveyed will be a combination of intentional and unintentional signs, i.e. a mix of impressions given and given off and the whole performance in question will be taken in, equally intentionally and unintentionally, in the context of each recipient's previous experience, assumptions, social settings, and psychological traits.

## On Game Theory

Game theory may be summed up thus: "*all actions a party assesses and opts to undertake or not, especially those actions it considers externally observable, integrate the existence of other parties and its understanding of the tactical options open to the other parties in question; naturally, the reverse is true too*"[157].

To operationalize this *strategic-exploration-of-own-options-on-the-basis-of-other-parties'-options-perceived-by-one*, game theorists manipulate discrete or continuous (and usually convex) functions.

[157] Tirole, J. 1988. Noncooperative Game Theory: A User's Manual *The Theory of Industrial Organization*: 479: The MIT Press.

## On Decision Theory

Indeed the field of decision theory[158] model-building rests on much the same premise as game theory. But just as knowledge is predominantly tacit rather than formal, so is individual and collective decision-making.

In the course of a given day, one makes all of one's individual decisions unconsciously. If a person were to ponder retroactively any such given decision she would be hard pressed to recollect exactly when it was made. On the other hand, she could easily define the time frame within which it was made; i.e. identify two moments, (a) a first moment clearly anterior to the decision being made (e.g. prior to the derived action ever being considered) and (b) a second moment clearly posterior to it being made (e.g. when the derived action was taken). The interval between these two moments can easily span hours and sometimes be longer.

In the case of collective decision-making, the social nature of the decision process blurs matters significantly more, even in the most formalized of processes. Take any investment committee within an organization. For any given investment, does the decisive moment occur when the committee greenlights the expenditure? Or should it be defined as the moment when the committee secretary schedules the investment decision on the committee agenda? Or perhaps it is the moment the investment case officer decides to run it by the committee? In some cases, committee deliberations encompass real decisions, particularly on touch-and-go cases. In most, though, the upstream process of prepping a case for committee socializes the investment decision. The investment committee merely acts formally as an acknowledging agent, rubber-stamping a decision already tacitly made. The interval between the two moments, the one prior to the collective decision tacitly being made, and the moment clearly posterior to it have been made, may thus span months, if not years.

---

[158] Simon, H. A. 1960. *The new science of management decision.* New York: Harper.

## On Reflexivity

In other words, game theory and impression management acknowledge the same thing: sentient parties are (a) aware of themselves, (b) aware of their surroundings and of other sentient parties, and (c) aware of the interactions between the latter and themselves. Game theorists express these dynamics as mathematical functions; interactionists, as narratives.

Hence the greater the gap between (i) perceptions by one party of another party's tactical options and (ii) the latter's real options, the noisier the input for the former's planning. Increased confrontation by involved parties between actual knowledge and other resources and their respective representations will attenuate this gap.

## On Cybernetics of Cybernetics

Second order cybernetics, also known as the cybernetics of cybernetics, examines the construction of models of cybernetic systems. It investigates cybernetics with the awareness that the investigators are part of the system, and factors in the problem of self-reference, self-organizing, and the subject-object problem.

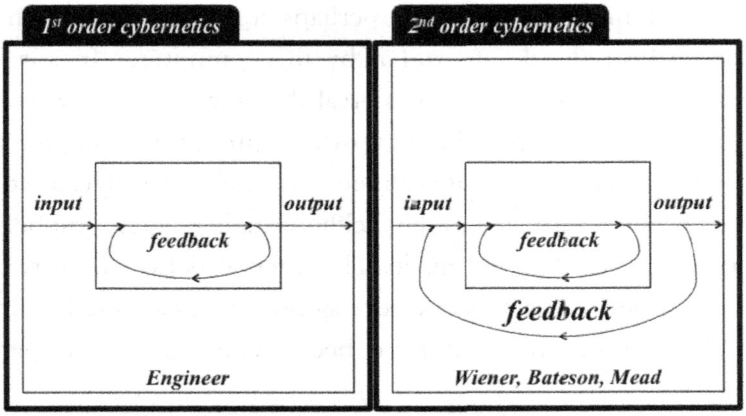


*Figure 17: Contrast between First and Second Order Cybernetics*

The diagram reproduced in Figure 17, by Mead and Bateson[159], contrasts first- and second-order Cybernetics. It emphasizes the requirement for a possibly constructivist participant observer in the second-order case.

Von Foerster[160] attributes the origin of second-order cybernetics to attempts by classical cyberneticians to construct a model of the mind. Researchers realized that "*a brain is required to write a theory of a brain; from this follows that a theory of the brain that has any aspirations for completeness has to account for the writing of this theory; and even more fascinating, the writer of this theory has to account for her or himself; translated into the domain of cybernetics, the cybernetician, by entering his own domain, has to account for his or her own activity: cybernetics then becomes cybernetics of cybernetics, or second-order cybernetics*".

## On Infinite Regress and Recursive Thought

On the one hand, Korzybski[161] introduces in his treatise on general semantics the statement that "*the map is not the territory*" inspired by modern advances in mathematics and physics; Bateson[162] deconstructs it in the context of cognition.

---

159 Mead, M., & Bateson, G. 1976. For God's Sake, Margaret, *CoEvolution Quarterly*. Menlo Park, CA: Whole Earth Access Catalog.

160 von Foerster, H. 2003. *Understanding Understanding: Essays on Cybernetics and Cognition*. New York: Springer-Verlag.

161 Korzybski, A. 1933. A Non-Aristotelian System and its Necessity for Rigour in Mathematics and Physics, *Science and Sanity*: 747–761.

162 Bateson, G. 2000. *Steps to an Ecology of Mind*. Chicago: University of Chicago Press, von Foerster, H. 2003. *Understanding Understanding: Essays on Cybernetics and Cognition*. New York: Springer-Verlag.

On the other hand, Weick[163], reflecting on the decade of organization theory development in the 1990s, warns not to underestimate "*the consequences that can follow if people are unable or unwilling to bound or voluntarily terminate their reflecting*", and observes that "*people who reflect on their own practices of theory development are currently torn between doing a shorter, shallower review, followed by update and resumption of their craft, and a longer, deeper review that may never reach updating or resumption*".

So while infinite regress carries the risk of thought paralysis, recursion, i.e. controlled regression towards a loop or a starting point, is a worthy approach.

## On Cognitive Socio-Psychology of the Media

In the media industry, consumers and producers of content do not meet[164]. The former create cognitive and affective representations of the latter and vice versa.

In effect, representations of consumers are virtually present at all stages of content production. With these representations, come judgments, attitudes, behaviors, beliefs, and knowledge regarding seldom-encountered consumers. Hence, the reliance of the industry —or the portion of industry professionals aware of the presence of audiences— on marketing tools (surveys, focus groups, sneak previews, etc.) to better discern audiences. Naturally, another portion of content producers are simply expressing themselves regardless of whether anyone is watching.

The situation is asymmetric. Cognitive representations by media consumers of media contributors are tempered by the fact that media contributors conduct impression-management efforts more professionally, whereas media consumers engage in little impression-management on media contributors.

---

163 Weick, K. E. 1999. Theory construction as disciplined reflexivity: tradeoffs in the 90s. *Academy of Management Review*, 24(4): 797-806.

164 Chabrol, C., Courbet, D., & Fourquet-Courbet, M.-P. 2004. *Psychologie sociale, traitements et effets des médias*, Vol. 2008.

However, where media is consumer-generated, i.e. where any consumer may contribute, content flows bi-directionally and on- and off-line social networking mechanisms engage, albeit not solely between individuals but also between individuals and the cognitive representations individuals form of (a) other individuals they haven't met but know of and (b) groups they aren't part of but are aware of.

## The Math is Not the Territory

When weighing past concepts with present and prospective findings, social scientists, being well versed in the epistemological, cognitive socio-psychological and management theories cited herein, are adept at questioning the tenets of the body of theories in their respective disciplines, because their daily pursuits and the underlying postulates are in close cognitive propinquity.

However, they are more likely to apply dogmatically and unquestioningly postulates that are cognitively removed from their discipline, like the mathematical tools they use to shape or test their theories. This shortcoming should be most blatant in the application of formal predicate logic, number theory, set theory, group theory and the like, (a) whose postulates even mathematicians must deliberately marshal their abstract thought processes to question and (b) whose applications, as part of primary and middle school curricula, are deeply embedded in our culture.

## The importance of being earnest —Social scientists are the instrument of social science

Social scientists, by the nature of their endeavors, should always keep in mind that, in their conversation with social phenomena, measure and instrument are irrevocably tied. Indeed, there can be no acknowledgement of a social phenomenon without an acknowledger, i.e. the social scientist himself.

Those social scientists who deny this nature and illusively seek a veneer of objectivity run on the contrary the risk of negating their research, by unraveling centuries of progress in pursuit of the aforementioned understanding of understanding. Conversely, since the discovery of quantum physics, the natural sciences have progressed towards this grasp of the intermingling of instrument and measure, which, though always present in the background of classical physics, stands at the forefront of contemporary physics.

## Quantum Leap

Piaget[165] states "*the three fundamental notions all social structures break down to are those of rules (or norms, from the subject's point of view), values and signs; and these appear at first glance not to be separable into concepts employed in natural science. … the distinction between (causal) explanation and understanding (of meanings and motives), however founded it may be, is in no fashion of a nature to oppose social and natural science: it reveals two irreducible albeit inseparable aspects of knowledge, and all science tends to reconcile them in varied manners, in the human subject that is their study object on the one hand, and in the coordination of experimental data from the study object and the deductive instruments provided by the subject on the other*".

Bitbol[166] further explains "*in quantum physics, what separates the explored object from the exploratory means meets fundamental limits; as in social science, quantum physics integrates into the to-be-known phenomenon the implications of knowledge gathering practices; because explored domain and means of exploration are tied, quantum physics cannot claim to* describe *as an outsider the properties of objects. Quantum physics specifies the evolution of a probabilistic prediction instrument (the quantum state vector). As in social science, quantum studies pore over* potentialities *and no longer on supposedly instrument-independent* actualities".

---

165 Piaget, J. 1967. *Logique et connaissance scientifique*. Paris: Encyclopédie de la Pléiade.
166 Bitbol, M. 2010. Théorie quantique et sciences humaines — Les savoirs du non-séparé.

## Copernicus's Lesson

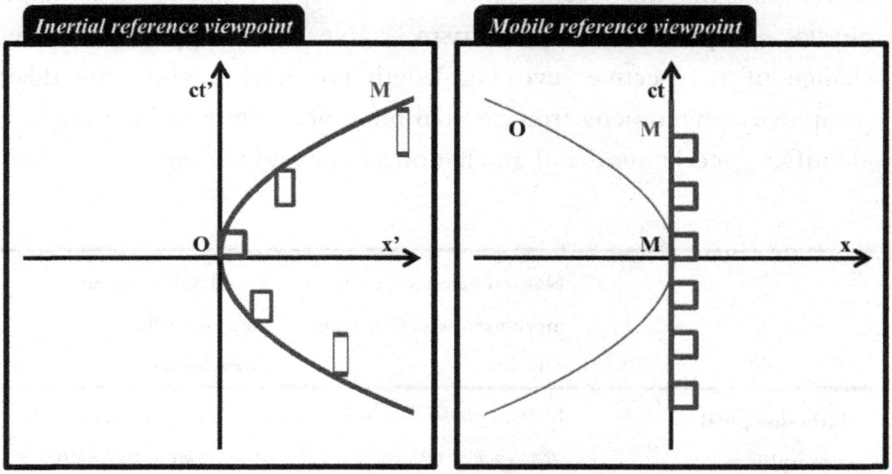


*initial conditions*: $x = x' = 0, \quad t = t' = 0, \quad v = 0$

*mobile M acceleration is A*

$$\Rightarrow v = \frac{At}{\sqrt{1+\left(\frac{At}{c}\right)^2}}$$

$$\Rightarrow x = \frac{c^2}{A}\left(\sqrt{1+\left(\frac{At}{c}\right)^2} - 1\right) \Leftrightarrow \left(\frac{Ax}{c^2}+1\right)^2 - \left(\frac{Act}{c^2}\right)^2 = 1$$

*Length contraction* *Time dilation*

$$ct' = ct\sqrt{1-\left(\frac{v}{c}\right)^2} \qquad x' = \frac{x}{\sqrt{1-\left(\frac{v}{c}\right)^2}}$$

$$at\ v = .4c, \quad \sqrt{1-\left(\tfrac{v}{c}\right)^2} < .917$$

*which means relativistic effects become noticeable*

*Figure 18: Einsteinian contrast between two perspectives of a mobile frame under constant acceleration*[167]

---

167 Semay, C., & Silvestre-Brac, B. 2005. *Relativité restreinte : Bases et applications.* Paris: Dunod.

A scientific breakthrough often occurs when a change of perspective reveals insights that seem natural in retrospect. Examples abound in physics: Copernican heliocentrism[168], Newton's apple[169], Einstein's change of perspective revealing length contraction and time dilation complexify phenomena from inertial reference points or his later insight identifying acceleration and gravitation as one and the same.

| | **Natural Sciences, and, more narrowly, Quantum Physics** | **Social Sciences, and, more broadly, Humanities** |
|---|---|---|
| **Methodological isomorphism** | In microphysics, *research strategies are formalized* because 'what [physically] is' (or what is studied) is *inseparable* from the approach employed to study it. | In social sciences, *research strategies are formalized* to *legitimize* epistemological advances. |
| **Dilthey's methodological distinction** | Practitioners of Natural Sciences *contemplate from a distance* their study object and articulate their findings through laws, i.e. *explain.* | Practitioners of Humanities *project themselves in the place* of their study subjects' situations and fathom their predicament, i.e. *understand.* |

168 Copernicus, N. 1543. *De revolutionibus orbium coelestium.*

169 Stukeley, W. 1752. Memoirs of Sir Isaac Newton's Life. London: Royal Society.

<table>
<tr><th></th><th>Natural Sciences, and, more narrowly, Quantum Physics</th><th>Social Sciences, and, more broadly, Humanities</th></tr>
<tr><td rowspan="2">**Habermas's methodological distinction**</td><td colspan="2">"Natural science theories present themselves as statements on the state of things whereas in humanities, the complex link between statements and the state of things is already present in the state of things analyzed."</td></tr>
<tr><td></td><td>The manner in which state of things is reported upon has an effect on same state, a self-realizing effect</td></tr>
<tr><td>**Habermas's methodological distinction (citing van Kempski)**</td><td>"...theoretical physics refer to *real* nature."</td><td>"Social Science studies are fundamentally studies of *possible* actions, whereas..."</td></tr>
<tr><td rowspan="2">**Piaget**</td><td>Natural sciences deal with *causes* from the perspective of the studied objects and with *reasons* from the perspective of the studying subjects that are organizing their linkages.</td><td>Social structures may be reduced to *rules*, *values* and *signs*; all three seem irreducible.</td></tr>
<tr><td colspan="2">"The distinction between the explanation (of causes) and the understanding (of signs and intents), however founded, is not of a character to oppose social and natural sciences: it reveals two irreducible but "indissociable" aspects of knowledge, and all science tends to conciliate them in diverse fashions, the ones in the human subject which is their very study object, the others in the coordination between experimental data from the study object and the deductive instruments provided by the subject."</td></tr>
</table>

| | **Natural Sciences, and, more narrowly, Quantum Physics** | **Social Sciences, and, more broadly, Humanities** |
|---|---|---|
| **Enter Quantum Physics** | In quantum physics, the separation between what is explored and means of exploration meets fundamental limits. | As does social science, quantum physics takes into account the implications of knowledge practices in the phenomenon to be known. |
| | Due to the limits of separation between explored domain and means of exploration, quantum physics cannot claim to describe, as from without, the properties of its study objects. Quantum physics describes the evolution of a probabilistic predictive tool (the quantum state vector). | As does social science, quantum physics is the study of *potentialities* and no longer *actualities* supposedly independent of the measure instrument. |

*Table 5: Social Sciences & Natural Sciences contrasts, inspired by Bitbol*[170]

Quantum mechanics produce riddles and paradoxes, because prior to the emergence of quantum physics, we thought natural science described the world. The aim of quantum mechanics, however, is the inventory of our possibilities to the world, i.e. "*we have to remember that what we observe is not nature in itself but nature exposed to our method of questioning*"[171].

---

[170] Bitbol, M. 2010. *Théorie quantique et sciences humaines — Les savoirs du non-séparé.*

[171] Heisenberg, W. 1958. *Physics and Philosophy*. George Allen and Unwin Edition.

Simmel[172]'s questioning of the very emergence of society —and introduction of social networks—, Goffman[173]'s reframing of social interactions from expression to intent to impress —and introduction of impression management—, Nash[174]'s reframing of strategic options in non cooperative settings —and introduction of game theory—, are examples of changes of perspective in social science which led to new, seminal insights.

More recently, in genetics, Dawkins[175] reframing of living beings as merely the vessels by which genes perpetuate themselves is another radical reframing. In cognitive science, Schopman and Shawky[176]'s "*a neural network doesn't represent the outside world but rather represents the interaction process between environment and organism*" echoes Heisenberg's here-above quote in our field. And in management studies, Donaldson[177] draws from the observation that both organization theory and practice are grounded in statistically supported methodology that scrutiny of our methods and tools, and more specifically of where these fail us, brings insights which have generally been overlooked up to now —e.g. when the paucity of observed events leads scholars and managers to draw statistically unfounded conclusions. We miss the proverbial forest by focusing exclusively on the statistics we've collected on the trees.

---

172 Simmel, G. 1910-11. How is society possible? *The American Journal of Sociology*(16): 372-391.

173 Goffman, E. 1959. *The Presentation of Self in Everyday Life*: Anchor Books.

174 Nash, J. F. 1951. Non-Cooperative Games. *The Annals of Mathematics*, 54(2): 286-295.

175 Dawkins, R. 1989b. *The Selfish Gene*. Oxford: Oxford University Press.

176 Schopman, J., & Shawky, A. 1996. Remarks on the impact of connectionism on our thinking about concepts. In A. Clark, & P. J. R. Millican (Eds.), *Connectionism, Concepts and Folk Psychology - The Legacy of Alan Turing*, Vol. 2. Oxford: Oxford University Press.

177 Donaldson, L. 2010. *The Meta-Analytic Organization, introducing Statistico-Organizational Theory*. Armonk, NY: M.E Sharpe.

## Model assumptions, suggestions for improvement

### Model Assumptions in Social Sciences

Social science has long used discourse, rhetoric and logic in the written word to pursue arguments, develop methods, to conduct experiments and so on.

However, building mathematically supported models has now become a major part of social science research. The sophistication of these models is greatest in the areas of economics, finance, and business.

Modelers often assume:

(a) continuity where phenomena measured is essentially discrete —e.g. decision theory;

(b) information to travel instantly, and be absolute and universal, albeit asymmetric in some cases —i.e. just known only to some parties, and unknown by others[178];

(c) cognitive processes to understand unlimited precision —e.g. monetary theories ignore that agents (i.e. persons) neither grasp the nuance between a trillion and a quintillion, nor see a difference between 1,000,000,353 and 1,000,000,535;

(d) cognitive processes not to loop (much less doubly) with their milieu —i.e. sentient beings are self-aware, aware of their environment/milieu and of their interaction with said milieu[179];

(e) the behavior of their model to be… model —i.e. mathematics are a social construction and a work in progress[180] but the body of math modelers know is assumed to be the right math for their context.

---

[178] Akerlof, G. A. 1970. The Market for "Lemons": Quality Uncertainty and the Market Mechanism. *The Quarterly Journal of Economics*, 84(3): 488-500.

[179] Goffman, E. 1959. *The Presentation of Self in Everyday Life*: Anchor Books.

Tirole, J. 1988. Noncooperative Game Theory: A User's Manual *The Theory of Industrial Organization*: 479: The MIT Press.

[180] Dehornoy, P. 2009. Cantor et les infinis. In S. M. d. France (Ed.), *Un texte, un mathématicien*.

## Recommendations for Cognitive Modeling

Modelers in economics, finance, business and associated fields assume the math of their day —and on which they were instructed, rather than educated, from primary school onwards— to be the right math for their model. I suggest *new algebras, new functions, new sets be built to fit social phenomena,* just as theoretical physicists built new algebras from raw, natural phenomena.

Phenomena are mostly discrete; measures always are. I suggest *the use of Robinson, rather than Peano, arithmetic,* unless induction can be explicitly justified in the model's context. Everyone cognitively understands 1, understands 2, and understands the difference between 1 and 2, but understanding the difference between 1,000,000,004 and 1,000,000,003 does not inductively follow. Whether we can reach all 'social' numbers by starting at 1 (or 0) and incrementing is doubtful in practice, given our cognitive and experiential boundaries.

Information in social science is never independent of the processes that bring it to light, transform it, transport it (at speeds slower than that of light), combine, share, or interpret it; nor is it independent of the actors involved in these processes. I suggest *applying a version of quantum information theory,* including light cones and event horizons, to model the social nature of information more closely.

Because infinitely precise cognition is a false assumption, I suggest acknowledging the feedback loops which are known to regulate cognitive processes, and doing away with non-looping cognitive models and *developing a body of theory, an algebra of recursive functions for use in cognitive models.*

However, infinitely looping models such as Mandelbrot[181]'s fractals ignore bounded rationality: I suggest these *cognitive recursive functions be finitely recursive* and introduce *bounded recursion.* Bounded recursion may be defined as recursion that loops back to itself after a limited number of steps (say under 10, 20 or 40 perhaps).

Lastly, I suggest embedding in this 'social' number theory algebra *explicit manifestations of the implied numerical collection methods and inherent errors,* as inspired by statistico-organizational theory[182,183].

## To Infinity and Beyond

Finally as a check on the modelers' hubris, I suggest their training include the following, mathematically proven[184] caution: As codified, our power of expression is countable; thus, the range of our potential expression is countably infinite. On the other hand, the range of phenomena we experience (and that we aspire to model) is uncountably infinite. Therefore, the knowledge we build may make significant advances, as we increase our power of expression when we coin new concepts. However, it does not decrease what remains uncharted, which is of a higher order of infinity. Models, as formal exercises, lack the intensity only tacit knowledge can provide[185].

---

181 Mandelbrot, B. B., & Hudson, R. L. 2004. *The (mis)behavior of markets: a fractal view of risk, ruin, and reward.* New York: Published by Basic Books.

182 Donaldson, L. 2010. *The Meta-Analytic Organization, introducing Statistico-Organizational Theory.* Armonk, NY: M.E Sharpe.

183 I.e. instead of expressing an aggregate sum and average of five samples as resp. 100 and 20 where the measures are 25, 19, 30, 10, 16 whose individual margin of error is 5, let's express the measures explicitly as $25_1 \pm 5$, $19_1 \pm 5$, $30_1 \pm 5$, $10_1 \pm 5$, $16_1 \pm 5$, and the sum and average as $100_5 \pm 25$ and $20_5 \pm 5$ or, alternatively as $[20,30]_1$, $[14,24]_1$, $[25,35]_1$, $[5,15]_1$, $[11,21]_1$ and, $[75,125]_5$ and $[20,30]_5$.

184 Cantor, G. 1874. Ueber eine Eigenschaft des Inbegriffes aller reellen algebraischen Zahlen. *Journal für die reine und angewandte Mathematik*, 77: 258-262.

185 Polanyi, M. 1966. *The Tacit Dimension.* London: Routledge & Kegan Paul.

## The next stage: deconstructing Boolean logic ?

I have hereabove questioned several tenets of mathematics that few who apply mathematics do, including number theory and geometry, which are usually accepted as universal and applied without question. However, though I could have, pulled the rug out from under the principle of Contradiction[186], which rules Western ("our") logic, I have not done so.

A case can be made for forgoing Western logic in cultural contexts that have developed independently of Western thought —e.g. in China. Table 8 tentatively illustrates where this might lead, albeit without delving into the well-explored concept of unknowable truths[187]. This author has enough trouble wrapping his head around Chinese space-time[188]. Ripping (admittedly Western in tradition) thought patterns away towards a non-Aristotelian system has been cautiously kept beyond the scope of this thesis.

| p | ¬p |
|---|---|
| p=T | F |
| p=F | T |

| p x q | q=T | q=F |
|---|---|---|
| p=T | T | F |
| p=F | F | F |

| p + q | q=T | q=F |
|---|---|---|
| p=T | T | T |
| p=F | T | F |

*Table 6: Boolean logic*

| p | ¬p |
|---|---|
| P=T | F |
| P=F | T |
| P=U | U |

| p x q | q=T | q=F | q=U |
|---|---|---|---|
| p=T | T | F | U |
| p=F | F | F | F |
| p=U | U | F | U |

| p + q | q=T | q=F | Q=U |
|---|---|---|---|
| p=T | T | T | T |
| p=F | T | F | U |
| p=U | T | U | U |

*Table 7: Extension to Kleene (ternary) logic, used in quantum computing*

---

186 Aristotle. 2000b. *Métaphysique* (J. Tricot, Trans.). Paris: Vrin.

187 Cook, R. T. 2006. Knights, knaves and unknowable truths. *Analysis*, 66(289): 10-16.

188 Granet, M. 1934. *La pensée chinoise*. Paris: Albin Michel, Jullien, F. 2009. *L'invention de l'idéal et le destin de l'Europe*. Paris: Seuil.

| P | ¬p |
|---|---|
| p=T | F |
| p=F | T |
| p=U | U |
| p=Ŧ | Ŧ |

| p x q | q=T | q=F | q=U | q=Ŧ |
|---|---|---|---|---|
| p=T | T | F | U | Ŧ |
| p=F | F | F | F | F |
| p=U | U | F | U | U |
| p=Ŧ | Ŧ | F | U | Ŧ |

| p + q | q=T | q=F | q=U | q=Ŧ |
|---|---|---|---|---|
| p=T | T | T | T | Ŧ |
| p=F | T | F | U | Ŧ |
| p=U | T | U | U | U |
| p=B | T | Ŧ | U | Ŧ |

*Table 8: Proposed extension to non-Aristotelian logic*

| | |
|---|---|
| **p, q** | propositions |
| **¬p** | opposite of **p** |
| **p x q** | **p** and **q** |
| **p + q** | **p** or **q** |
| **F** | false |
| **T** | true |
| **U** | unknown |
| **Ŧ** | true, false harmoniously entwined |

*Table 6, Table 7and Table 8 key*

# VII] AN APPLICATION FOR QUALS — IN THE *I* OF THE ENTREPRENEUR

In the context of a research workshop conducted by Isabelle Micaelli, I realized that entrepreneurs, and notably innovative venture creators, need to cognitively juggle between professional disciplines (and, perhaps, between national cultures, too).

The following is an extract I wrote alone of a paper we co-developed[189], which we hope will reveal, in conjunction with current methods[190], new insights for qualitative research method scholars.

This methodology thus proposed aims to enhance the current qualitative research tools range. Trials are scheduled beyond the scope of this thesis.

**Semiotic relations methodology**

Key words and expressions emerge from the innovative venture creator's self-narratives. The mere processes (a) of innovation, (b) of venture creation, and (c) of software research all concur for different reasons to situate the actor of these processes to span disciplines: resp. (a) design and usability, R&D, marketing; (b) law, management, sales, and finance; (c) engineering, design…

On the sample scrutinized, all innovative venture creators have encountered early in their life various national cultures.

This double realization (the numerous disciplinary and cultural paths crossed) invites us to explore what the key words and expressions carry in the various national cultures and professional disciplines encountered.

---

[189] Idelson, M., & Micaëlli, I. 2010. In the "I" of the innovative entrepreneur: a proposed qualitative analysis. In M. Graham, & Y. Li (Eds.), *CMS Organization workshop: Innovation and entrepreneurship track*. Montréal.

[190] Demazière, D., & Dubar, C. 1997. *Analyser les entretiens biographiques. L'exemple des récits d'insertion.* . Paris: Nathan.

Once these key words and expressions have emerged from the qualitative analysis of the self-narratives, we propose to map them into the following models.

## Semantic network mapping

For each key word and expression, we will map, in both the language of the self-narrative and the narrator's mother tongues, all potential meanings of each key word and expression (to learn of the potential polysemous contrasts in each national language). We will also categorize these within given disciplines if relevant (e.g. the English words *template*, *design*, *code* have a specific meaning in software).

In this 2-mode network, the nodes are cultures (national or professional), meanings (dually tacit and formal, irrespective of grammar —i.e. verbs and substantives are amalgamated) and expressions (including single words).

The relationships are (a) meaning in use in culture, (b) meaning expressed by word, and (c) word in use in culture. These three relationships are theoretically redundant. Maintaining the three does however bring flexibility to future analysis. A measure of frequency of use may qualify these relationships.

## Culture network mapping

For each language, dialect and professional jargon that emerged in map A, we will map their relative proximity to each other.

In this 1-mode network, the nodes are cultures and sub-cultures, proxied as before through language and vocabulary.

The relationships are (a) culture subsumed to other culture, (b) culture shares common root with other culture, (c) cultural works translated from other culture, and (d) culture has trade with other culture. The first relationship is binary, hierarchal and incomplete[191]. A measure of age of common root (e.g. 2500 BP) may qualify the second relationship[192]. Cultural[193] and trade[194] exchanges measures will qualify the two last relationships proposed.

## Semiotic relations methodology challenges

As with any network analysis, completeness of data is a challenge. The omission of one network node can radically alter the topology of a network, rendering a premature analysis moot.

Specific attention must be made to colloquial expressions that may arise in a subculture and that the researcher may miss, misinterpret or over interpret. As any single researcher cannot immerse herself in all cultures and subcultures, other members of the studied subculture not directly involved with the innovative venture creator should be sought out, and the symbolism and meaning of earmarked expressions should be crosschecked to avoid potential omissions, misinterpretations and over-interpretations. As much as possible, native speakers should be interviewed, not translators or interpreters, as the latter often unconsciously add detrimental bias and moderation to the exercise. Bias and moderation here are detrimental as they shield non-native speakers of the rawness of some expressions. Although this is welcome in diplomatic or negotiation contexts —i.e. the interpreter's natural habit-at—, it dulls this category of academic pursuit.

---

[191] International Standards Organisation. 2010. ISO 639-3 standard: SIL International.

[192] Gray, R. D., & Atkinson, Q. D. 2003. Language-tree divergence times support the Anatolian theory of Indo-European origin. *Nature*, 426: 435-439.

[193] Unesco. 2010. Translatium database: Unesco.

[194] World Trade Organization. 2010. WTO statistics database.

# VIII] IMPLICATIONS FOR PRACTICE — WHAT YOU THINK YOU KNOW

**v the case of the jargon drift**

A training institute, dedicated to the audiovisual media industry, was mandated to develop a curriculum for a satellite broadcaster's engineers and technicians on a new high definition television standard. Despite the standard having been well engineered and defined, developing the curriculum proved exceedingly arduous because the authors of the various elements of the standards used the same terminology to describe different concepts.

The authors of the video element of the standard used the term High Frequency Signal to describe the range of the video standard and Low Frequency to describe and dismiss anything to do with the audio standard —which they weren't interested in.

The authors of the audio element of the standard used the term High Frequency to describe anything to do with the audio standard and Low Frequency to describe and dismiss anything beyond their range.

Finally, to the satellite operators, the term High Frequency Signal described the audiovisual signal, carried according to the compounded standard.

What is striking in this context is that all these engineers were office neighbors, lunched daily with each other, but never talked shop. In effect, a community of practice, which had authored the original standard, through division of labor and topic specialization, evolved and split into several, each with mutually contradictory professional terms of reference.

Traditionally in German business education, doctoral degrees were prepared in both law and economics. Likewise, the case that has been presented calls for more explicit expressions of the core tenets of these disciplines that shape business activity, i.e. more, not less understanding of rhetoric, semantics, logic and mathematics in managerial education. From these and the explicit demonstration of the socially constructed nature of these disciplines, however counterintuitive this appears, one can hope managers will neither confuse the narrative with the story, nor the math with the territory.

**"The question was never raised!"**

The potential managerial implications this research sheds light on span both behavioral and cognitive-inspired guidelines at the cross-cultural and cross-disciplinary levels.

Contexts where these findings may bring progress to business practices include:

- Staffing, training and organization design when an individual or collective goal requires performance in a culturally or disciplinary novel context for a given organization, group or individual; e.g. (a) an engineering firm assigned to design a power plant in a new country, (b) a software development team assigned to design an information system for micro-credit processing, an activity it hasn't had experience with before, (c) an accountant assigned to explain for the first time movie production accounting standards to novices;

- Professionals undertaking modelling activities may gain insight from the underreported fact that mathematics are a social construction and the math, not just the economics they use are (self-evidently) the ones they know; however, they may not be the mathematics that are ultimately appropriate to the situation; this includes the number theory, set theory and algebra they were instructed upon in primary, middle and high school, and this mismatch may have implications in the context of regulation, governance, compliance, performance, design and measurement of organizations. In particular, all quantitative models leveraging arithmetic, algebra and analysis coincidental with their era may be improper to the object context being modelled;
- Similarly, quantitative models of social phenomena that ignore the bounded reflexivity that sentient cognition implies, will, when applied straightforwardly, engender perverse effects that model practitioners should be made explicitly aware of;
- Often implicit assumptions, including of consensus between cultures and disciplines of common terms of references, should be made explicit;
- The normative nature of vocabulary should be acknowledged and assumptions that two bodies using the same term mean the same thing should be regularly pondered;
- Professionals and managers endowed with a given skill set (fluency in a certain language, effectiveness at conducting certain tasks…) should, while exploiting their skill sets, take every opportunity to explore the limits of their knowledge and know-how.

Discovered lexical gaps between national cultures or professional disciplines that may be revealed may be symptoms of symbolic gaps, especially in the case of mathematical objects (from algebra, geometry, calculus…).

They may also be symptoms of semantic gaps, polysemy, e.g.:

- *accountability* in English has no counterpart in French;
- *certainty* in English, *certitude* in French have no counterpart in Chinese;
- English amalgamates the French *savoir* and *connaître* into *to know* though its Scottish cousin does mark the distinction in the use of *to ken* and *to wit*, two words with roots shared with Danish and German;
- simple words such as *find* for which one would expect implicit consensus can in practice be as context-driven as more sophisticated terms (see Figure 19 and Figure 20);
- the Globe study discovered in the field two different Greek words exist to convey *leader* in English (or *Führer* in German, another word that doesn't travel well) and use in the Greek version of its survey of *archondas* instead of *archon* distorted their initial results because the former term carries as subtext 'a rather arrogant upper class person'[195];

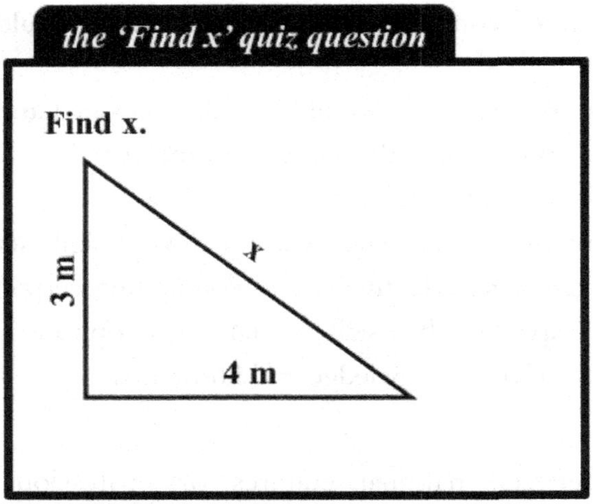


*Figure 19: The 'Find x' quiz question*

[195] House, R. J., Hanges, P. J., Javidan, M., Dorfman, P., & Gupta, V. 2006. *Culture, leadership, and organizations: the GLOBE study of 62 societies.* Thousand Oaks, CA: Sage.

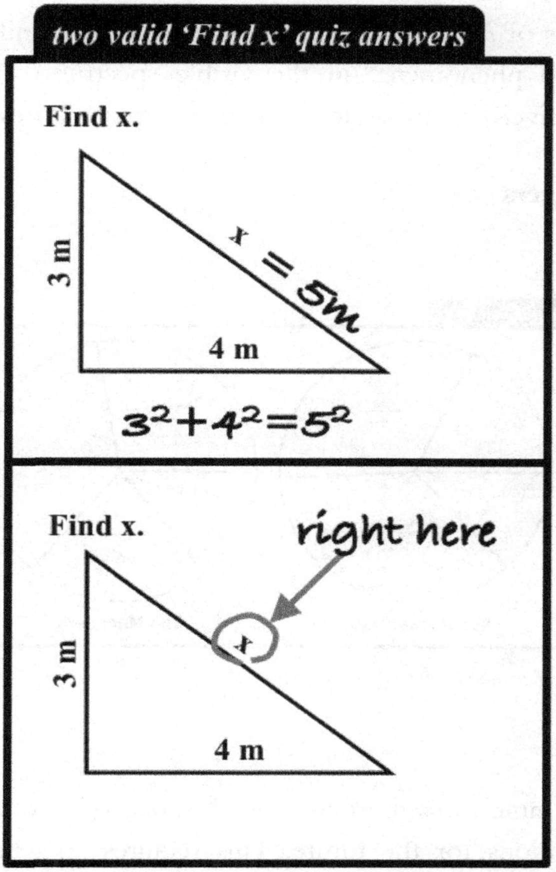


*Figure 20: Two valid 'Find x' answers, context-willing*

- organizational slack is perceived and understood radically differently by different schools of social scientists: to micro-economists, slack is negatively connoted and synonymous to inefficiency and waste; to organization scholars, slack conveys a positive impression, a feature of organizations that help them weather storms, a buffer ensuring flexibility and agility in shifting contexts;

- even the concepts of *blue* and *green*, which are arguably manifestations of physical phenomena (in the visible spectrum), have turned out to be diversely ontologically defined across cultures[196].

**Implications for managers**

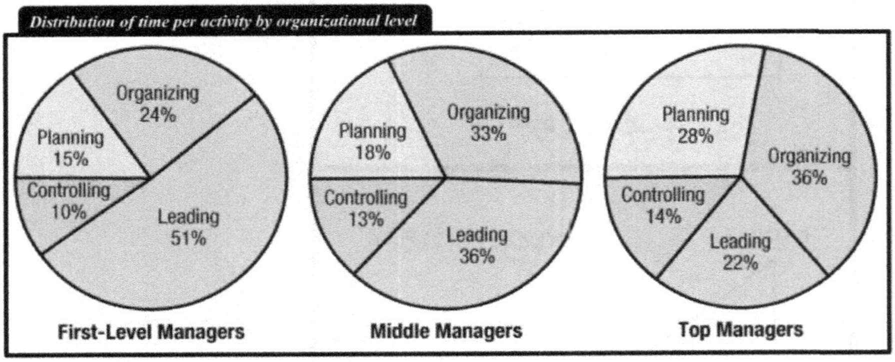


*Figure 21: Distribution of time per activity by organizational level* [197]

Top managers within large organizations spend most of the time preparing their organizations for the future. This requires (a) a broad understanding of where they expect their organization to go, (b) the wherewithal to surround themselves with staff adequate to the task and, last but not least, (c) the ability to communicate their vision to stakeholders to whom they are not directly connected using vocabulary that is an appropriate projection of their organization's identity.

---

196 Berry, J. W., Poortinga, Y. H., Segall, M. H., & Dasen, P. R. 2002. *Cross-cultural psychology: research and applications* (2nd ed.): Cambridge University Press.

197 Mahoney, T. A., Jerdee, T. H., & Carroll, S. J. 1965. The jobs of management. *Industrial Relations*, 4(2): 97–110.

To prevent misunderstandings that may mar planning activities and designing organizations, top managers therefore need (a) support staff collectively cognizant in all social representations (and their representative vocabulary) of the external and internal stakeholders involved within their remit and (b) middle management they can delegate to equally cognizant in all social representations (and their representative vocabulary) of the external and internal stakeholders involved within the more limited sphere of middle-management activity.

Any strategic-policy modeling may gain insight from making certain the underlying math and economics selected are appropriate to the object situation being modelled.

Middle managers staffing multi-disciplinary or multi-cultural teams may benefit from insuring the presence of one or several go-betweens

Middle managers staffing mono-disciplinary teams of professionals (e.g. teams of economists, teams of lawyers…) may find the teams' performance[198] improves if team members do not share the same theoretical grounding (e.g. different macroeconomics schools of thought, different legal traditions…), especially if the team's tasks involve creativity.

## Implications for corporate HR and corporate communications

Corporate (resp. divisional) HR or communications divisions often sponsor and maintain repositories of skill sets, business terms of reference, etc. that reflect the corporate (resp. divisional sub-) culture. Often in multinationals (MNCs), these repositories are multilingual.

---

198 For more on performance evaluation design and the qualitative, quantitative (continuous and discrete, ordinal or algebraic, …) nature of performance indicators, cf. Mangin, N. 2008. A comprehensive analytical framework for studying performance evaluation, *2008 Academy of Management Annual Meeting.* Anaheim, CA: Academy of Management.

However, especially, when the MNC's culture strongly echoes engineering cultural practices or the MNC has bought an off-the-shelf package designed solely by engineers, the approach undertaken in the population of these repositories is often heavily positivist indeed, as the World-Wide Web Consortium's semantic web initiative attests[199].

A major drawback is that several types of misunderstandings[200] can occur, e.g.:

- a term with a single meaning of in one language can mean several things in another;
- one meaning of a term in one language has no equivalent in another;
- one meaning of a term in one language is only approximately capture by its translation in another (though nothing in the corporate repository acknowledges this);
- a term used in one discipline may fork into two different, overlapping or distinct terms in another;
- a term used in two disciplines may mean radically different things;
- a term used by two teams or organizations may mean different things with insiders never confronting the contrast and outsiders erroneously drawing conclusions from generalities.

Not all misunderstandings have consequences, and the consequences that do occur vary in impact. However, designing repositories and open processes with slack and the leeway for partial translation can provide those that rely on them with the benefit of the doubt.

HR professionals can also usefully assist managers in specifying their staffing needs in light of this thesis' expected insights[201].

---

199 W3C. 1989. Semantic Web, Vol. 2010: World Wide Web Consortium.

200 For more on the sociolinguisitics of misunderstandings, cf. Labov, W. 2009. Volume 3: Cognitive and Cultural Factors, *Principles of Linguistic Change.*

201 HR professionals may also find insights in Wurtz, O. 2009. The impact of

## Implications for entrepreneurs

The central function of an entrepreneur is to move resources, however painfully, to areas where they can be used more productively[202]. From the social networks perspective, entrepreneurs are *tertius iungens*[203], explorers who identify gaps and move to exploit them as brokers in one fashion or another.

As leaders of structures nimbler than larger organizations, entrepreneurs need the same understanding of their stakeholders without necessarily having the comfort of support staff. On the other hand, concentration of skill sets into fewer go-betweens should enable entrepreneurs to thrive in environments where terms of reference shift rapidly, ensuring little direct competition from larger, less flexible firms.

## Implications for change management

Change management is seldom limited in scope to one discipline and one culture. The need to explicit implicit assumptions, to develop consensual terms of reference between cultures and disciplines or, conversely, to agree to disagree on some terms used and explicitly limit the scope of each term meaning to its domain should not be underestimated.

Section IX develops how post-merger integration could improve, given these findings.

---

personality on expatriate coping and cross-cultural adjustment, *2009 Academy of Management Annual Meeting*. Chicago, IL: Academy of Management.

202 Schumpeter, J. A. 1962. *Capitalism, Socialism, and Democracy* (3rd ed.). New York: Harper & Row.

203 Obstfeld, D. 2005. Social Networks, the Tertius Iungens Orientation, and Involvement in Innovation. *Administrative Science Quarterly*, 50(1): 100.

## Implications for corporate IT

Along with supporting top management, business lines, support functions and change programs, corporate IT is responsible for the smooth (or not so smooth) introduction of technology into organizations (bottom-up change spurred by technological ruptures). This can introduce new vocabulary too, and the potential for lexical misunderstanding.

## Implications for corporate governance

Staffing a board with directors savvy in the cultural and disciplinary vocabulary of the organization to be steered may seem self-evident but successfully achieving that task may be harder that staffing lower rungs in the organization, as executive search practices have clamored about the demographically shallow talent pool. Perhaps, as the tail end of baby-boomers retires in Western Europe and North America, will we witness a regional trend reversal.

## Implications for other stakeholders

Naturally, the implications hereabove may apply indirectly to (a) bodies that aim to shape and regulate the abovementioned professions (legislators, public policy think tanks, public regulators, self-regulating bodies…), (b) consulting firms serving the abovementioned professions as well as (c) professional education bodies (professional trade associations, business, law, economics and other technical academic institutions…) catering to the same.

## The value of self-doubt and open-mindedness

In a nutshell, operatives and managers in cross-disciplinary or cross-cultural settings can only be more effective if they doubt their knowledge in untried contexts and develop open-minded attitudes to explore the limits of their knowledge in question.

Hence, I can derive the following from my experience critically questionning my own West-grounded faith in numbers:

> The more self-doubt and open-mindedness exercised in a cross-cultural or cross-disciplinary context, the greater an operative's effectiveness.
>
> The more aware of cross-cultural and cross-disciplinary lexical gaps, the greater a manager's planning, organizing, leading and controlling effectiveness.

Measures to induce self-doubt and open-mindedness can include sensitivity training prior to cross-cultural or cross-disciplinary assignments.

Screening for self-doubt and open-mindedness can be integrated into staffing decisions for cross-cultural or cross-disciplinary assignments.

Measures to cultivate self-doubt and open-mindedness can include debriefing at the conclusion of cross-cultural or cross-disciplinary assignments and critical reviews of past performance.

# IX] AN APPLICATION TO PRACTICE — HOW WE CAN LEARN WHAT YOU AND I KNOW

"All mankind is of one author, and is one volume; when one man dies, one chapter is not torn out of the book, but translated into a better language; and every chapter must be so translated … No man is an island, entire of itself … any man's death diminishes me, because I am involved in mankind; and therefore never send to know for whom the bell tolls; it tolls for thee."[204]

**vi the case of the hapless investment banking merger**[205]

Contemporary merger execution is often a pressing exercise, especially in service industries. Chris, an IT director from Banking Union Corp (BUC), summed the process up as the *Mongol horde* approach to merging, "the guys from Allied Reserves Group (ARG) are now on hill A, we're on hill B and, in the next 5 weeks we'll become Banking Union Group (BUG), i.e. define hill C, towards which all of us will now rush and then sort out the cracks in our operations there".

In the course of preparing the merger (defining hill C), representatives of each business line from ARG and BUC met and described to one another their internal organization.

[204] John Donne Meditation XVII in Devotions upon Emergent Occasions (1623)

[205] In a sense, this case illustrates the social nature of commensuration put forward by Espeland, W. N., & Stevens, M. L. 1998. Commensuration as a Social Process. *Annual Review of Sociology*, 24: 313-343.

When the teams from ARG and BUC peddling to corporate customers Contracts for Difference (CFDs), a relatively unregulated financial derivative, met to plan their merger, they discovered the legal foundation of their offerings was so radically different it rendered any comparison between ARG CFDs and BUC CFDs moot.
This was a shock to executives on both sides as they had up to then competitively benchmarked their teams against each other and financial analysts had wantonly compared all CFD-providing units within investment banks with obvious disregard to the true legal underpinnings of the various offerings. This is all the more unfortunate since most financial analysts themselves sit within investment banks.

## Reflections on Information representation: zooming out of context

### On the Social Nature of Information ... Systems

The social nature of information has been raised in Section VI and that all individuals are a medium was observed.

A similar phenomenon occurs in information systems. All Information systems are also vehicles for shared languages, information and contexts; however, Information systems not being sentient, they act more like knowledge extensions of individuals and groups as they themselves lack the ability to interpret in context their information repository. A mark of good information system design therefore is that disparate groups of information system users interpret coherently the information shared within the information system in question. Often, this is achieved through effective dissemination of data about the data (metadata).

## When Information Systems collide

In the context of organizational mergers, merging Information systems represents a consequential proportion of the effort as well as a project lasting longer then the merger effort in both managerial discourse and mind. Figure 22[206] illustrates the overlapping organizations top management teams must cognitively superimpose during a merger process.

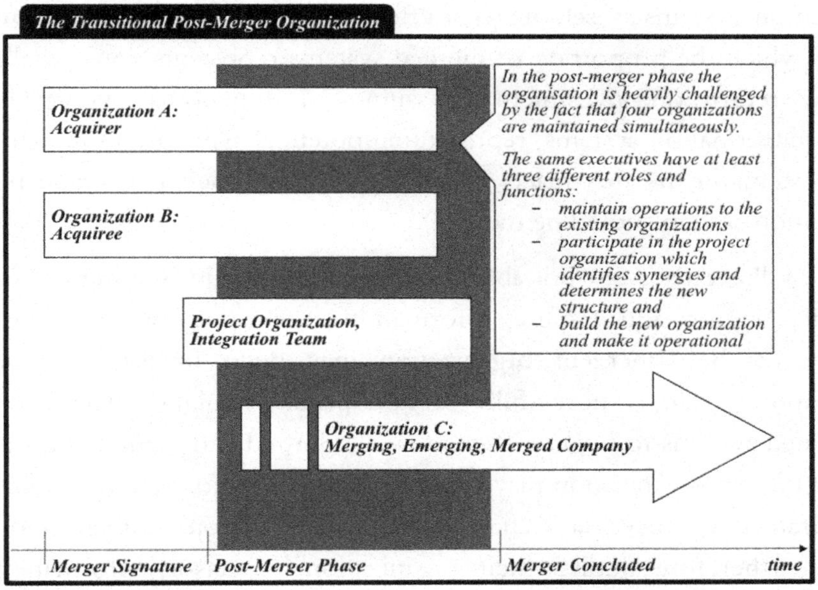


*Figure 22: The Transitional Post-Merger Organization*[207]

[206] Fendt, J. 2006. *The CEO in Post-Merger Situations: An Emerging Theory on the Management of Multiple Realities*. Leiden University, Leiden.

[207] Fendt (ibid.)

In practice, information system mergers outlast the formal 'Merger Concluded' milestone illustrated in Figure 22. Some Information systems take several more years to be merged or phased out; others are never merged at all and are merely sundowned, i.e. no new business is input into them and all existing supported business resides within this legacy information system until the business activity's natural run off.

This more leisurely pace of *rapprochement* is not necessarily a sign of inefficiency. On the contrary, a comprehensive merger of 100% of information systems is seldom cost-effective. There is a tipping point beyond which the proportion of merged systems represents a marginally negative return on investment —assuming a functional overlap in existing information systems, representing potential synergies in information systems in the first place (if not that tipping point is 0% and no information system should be merged).

Figure 23 illustrates a few possible outcomes for individuals systems. For example, Acquirer and Acquiree information systems may be retained or abandoned; the Emergent organization may build or acquire new information systems, successfully or not; plans to migrate data from abandoned systems to target systems may be drafted and carried out; the Merger Project organization may also be kitted with information systems whose fate is not necessarily tied to the transient Project organization and may therefore outlive their organizational sponsor; information system teams supporting all four organizations will also rely on information systems to carry their mission out.

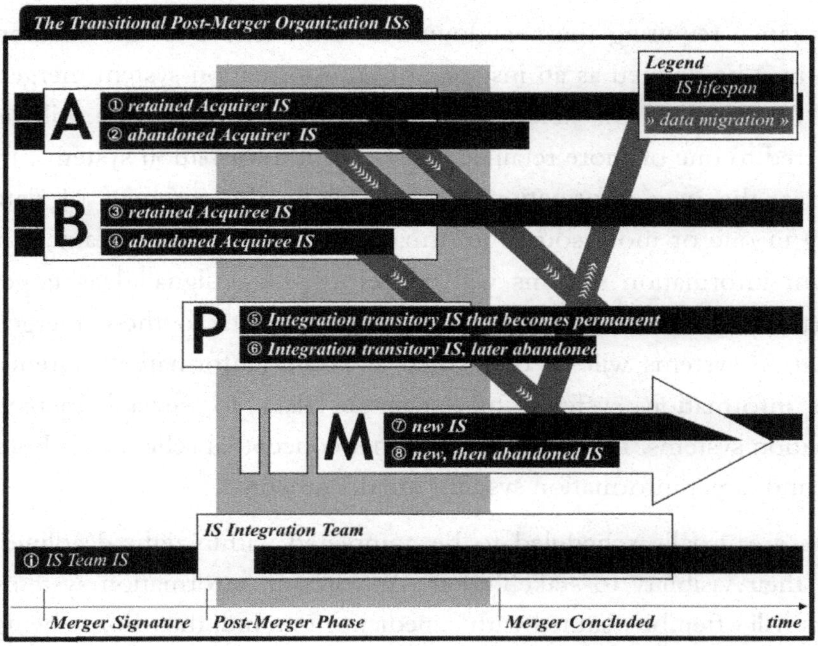


*Figure 23: The Transitional Post-Merger Organization Information System*[208]

[208] In this illustration:
Retained IS ① originates from dissipating Acquirer A, feeds data to transient Integration Project P IS ⑤, and is fed by coalescing Emerging Organization M IS ⑧.
Abandoned IS ② originates from A, feeds data to dissipating Acquiree B IS ③.
Retained IS ③ originates from B, and is fed by A IS ②.
Abandoned IS ④ originates from B, and feeds data to Emerging Organization M IS ⑦.
Transitory IS ⑤ is launched with P, is fed by A IS ①, and is perpetuated by M, thus outliving its sponsor P.
Transitory IS ⑥ is launched and terminated with P.
New IS ⑦ is launched with M, and is fed by B IS ④.
New IS ⑧ is launched with M, but abandoned, and feeds data to A IS ①.
Equivalent events occur within IS Team IS ⓘ originates which supports IS Integration team, whose lifespan exceeds P.

Any instance requiring the abandonment or launch of an information system may be defined as an instance of an information system merger. In the former case, the abandoned information system's data will be transferred to one or more retained or emergent information systems. In the latter, the new information system will be the recipient of data residing in one or more source information systems. These retained or emergent information systems will henceforth be designated as target information systems, and the sources that feed data to these merged information systems will be designated as source information systems. Source information systems may provide data to several merged information systems. It will be assumed that conceptual schemas of both source and target information systems are documented.

Mergers are usually scheduled to be completed within tight deadlines, due to their visibility to stakeholders. As a result, information systems mergers will often be planned with expediency in mind, the priority being to get the target organization "up and running" as soon as possible.

I will now propose a method to conduct information systems mergers that enables implicit contexts to be made explicit, and, provided the target information systems is able to accept all data fed into it from source information systems, ensures data migration is lossless, including in terms of traceability to source information systems.

## Modeling Information: database normalization

| | |
|---|---|
| **1st normal form 1NF**[209] | Table faithfully represents a relation and has no repeating groups. |
| **2nd normal form 2NF**[210] | No non-prime attribute in the table is functionally dependent on a proper subset of a candidate key. |
| **3rd normal form 3NF**[211] | Every non-prime attribute is non-transitively dependent on every candidate key in the table. |
| **Boyce-Codd normal form BCNF aka 3.5NF**[212] | Every non-trivial functional dependency in the table is a dependency on a superkey. |
| **4th normal form 4NF**[213] | Every non-trivial multivalued dependency in the table is a dependency on a superkey. |
| **5th normal form 5NF**[214] | Every non-trivial join dependency in the table is implied by the superkeys of the table. |
| **Domain/key normal form DKNF**[215] | Every constraint on the table is a logical consequence of the table's domain constraints and key constraints. |
| **6th normal form 6NF**[215] | Table features no non-trivial join dependencies at all (with reference to generalized join operator). |

*Table 9: The eight normal forms of database normalization*

[209] Codd, E. F. 1970. A Relational Model of Data for Large Shared Data Banks. *Communications of the ACM*, 13(6): 377–387.

[210] Codd, E. F. 1972. Further Normalization of the Data Base Relational Model. In R. J. Rustin (Ed.), *Data Base Systems: Courant Computer Science Symposia Series 6*: Prentice-Hall.

[211] Codd 1972, ibid.

[212] Codd, E. F. 1974. Recent Investigations into Relational Data Base Systems (IBM Research Report RJ1385), *Proc. 1974 Congress*. New York: North-Holland.

[213] Fagin, R. 1977. Multivalued Dependencies and a New Normal Form for Relational Databases. *ACM Transactions on Database Systems*, 2(1): 267.

[214] Fagin, R. 1979. Normal Forms and Relational Database Operators, *International Conference on Management of Data*. Boston: ACM SIGMOD.

[215] Fagin, R. 1981. A Normal Form for Relational Databases That Is Based on Domains and Keys. *Communications of the ACM*, 6(3): 387-415.

Information Systems (IS) practitioners have long used database normalization as a standard model for information systems. This process is generally expressed in the eight normal forms in Table 9.

| Name | Purpose | Sponsor [underlying format] |
|---|---|---|
| **Extended Business Reporting Language (XBRL)[217]** | Universal exchange of business communications | The XBRL consortium [XML] |
| **Extended Markup Language (XML)[218]** | Universal exchange of data | WorldWide Web Council [superset of HTML, i.e. extends HTML's power of expression beyond Human-Computer Interfaces] |
| **HyperText Markup Language (HTML)** | Web page display | W3C [plain text][219] |
| **Internet Calendar (iCalendar)** | Universal exchange of calendar events, to-do items, journal entries, free/busy time | Internet Mail Consortium [plain text][220] |
| **Keyhole Markup Language (KML)[221]** | Universal exchange of geographical data | Open Geospatial Consortium [XML] |
| **Real Simple Syndication (RSS)[222]** | Universal exchange of web feed syndication | Berkman Center for Internet & Society at Harvard Law School [XML] |

[216] Date, C. J., Darwen, H., & Lorentzos, N. A. 2003. *Temporal Data and the Relational Model: A Detailed Investigation into the Application of Interval and Relation Theory to the Problem of Temporal Database Management.* Oxford: Elsevier.
[217] XBRL Consortium. 2010. XBRL Specifications.
[218] W3C. 2008. XML.
[219] W3C. 1999. HTML.
[220] IMC. 1998. Personal Data Interchange.
[221] OGC. 2008. KML.
[222] Harvard Law. 2003. RSS.

| Name | Purpose | Sponsor [underlying format] |
| --- | --- | --- |
| **Visiting Card (vCard)**[223] | Universal exchange of business cards | Internet Mail Consortium [plain text] |
| **Wireless Markup Language (WML)**[224] | Augmented web page display on mobile phones | Open Mobile Alliance [XML] |

*Table 10: Internet era file exchange format de facto standards*

Though computers (and information systems) have been networked since the 1960s, information system to information system interface building had been clunky and unwieldy until the advent of the World-Wide Web Consortium-engineered eXtended Markup Language (XML) and sundry microformats, on the coattails of the Internet. Nowadays, information system to information system exchanges are commonly XML-based, unless network bandwidth issues are a concern (XML is notoriously verbose, adding sometimes as much a multiplier overhead).

While the organization of system-to-system exchanges has been evolving, a parallel evolution has been going on in the internal organization of data within information systems. Trends in this area have been threefold:

(a) processing, memory and network resources have become more affordable;

(b) business needs (including functional, performance, uptime, and interoperability) have been evolving faster;

and, (c) information access and display options have mushroomed so much that information system architectures have attempted to completely decouple device delivery of information from the underlying storage and data management systems.

223 IMC. 1998. Personal Data Interchange.

224 OMA. 2000. WML.

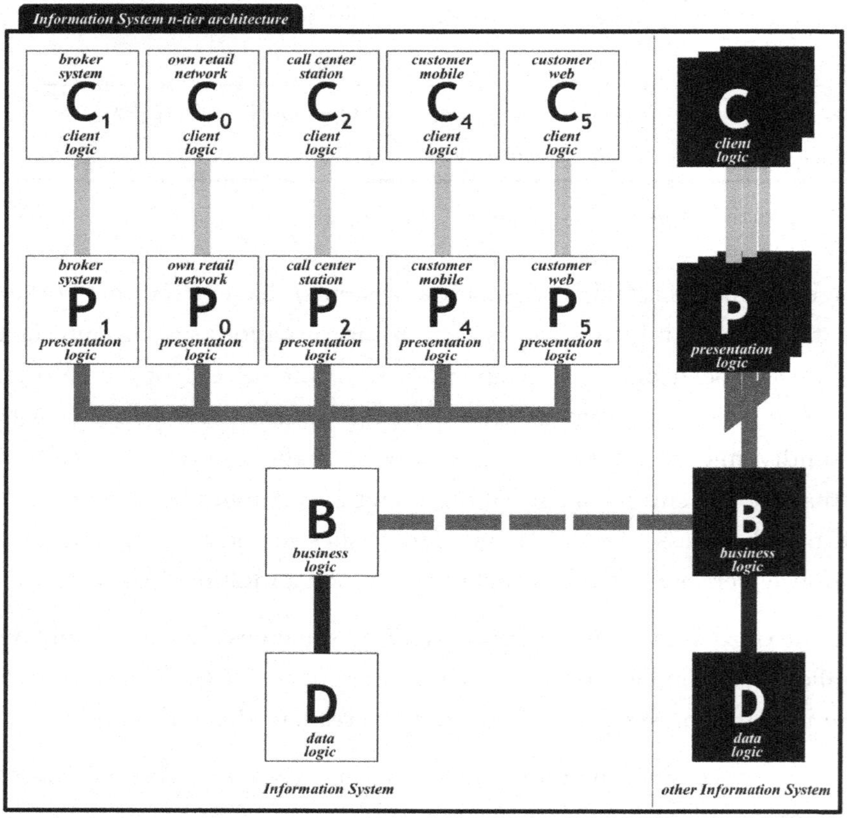


*Figure 24: A typical Internet-era n-tier information system architecture*

This has led to the development of n-tier architectures, as illustrated in Figure 24, which enables loosely coupled application-to-application service-oriented architectures.

This section will now focus on information system data logic.

## Three-level architecture

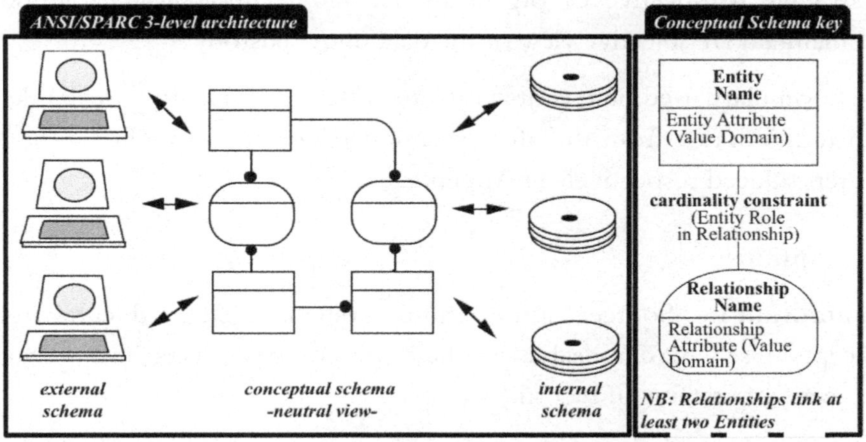


*Figure 25: Three-Schema Approach and Conceptual Schema Key*[225]

Information system data logic is generally agreed to be modeled on three different levels[226], as illustrated in Figure 25:

(a) external models (or schemas) of an information system, corresponding to a particular scenario within which the information system is to be used (e.g. data entry, interface with another information system, web display…),

(b) conceptual models and

(c) internal models of an information system corresponding to their actual implementation in a given computer or network of computers.

---

[225] National Institute of Standards and Technology. 1993. Federal Information Processing Standards Publication 184: Integration Definition for Information Modeling (IDEF1X).

[226] NIST, ibid.

The former tend to evolve according to organizational needs, the latter according to database management systems or other underlying technological evolution. Conceptual models are developed to maintain a neutral perspective irrespective of organizational and technological evolution, and maintain an objective view of the data under custody.

The vocabulary used here is drawn from IDEF1X's glossary[227], extended with terms drawn from the afore-mentioned normal forms literature[228], and reproduced respectively in Appendixes.

## Assumptions

**Assumption 1:** Source information systems have documented conceptual schemas traceable to their present custodians and tracing their present systems of record.

> Source information systems are assumed to be comprised of at least three elements:
>
> > (a) information pertaining to their current (pre-merger) respective custodians, i.e. their organizational sponsors,
> >
> > (b) their conceptual Data Schemas, comprised of entities, entity attributes, relationships, relationship attributes, and domains, in normalized form,
> >
> > (c) their present Systems of Record,
> >
> > and (d) documented correspondence between these Data Schemas and Systems of Record, source information system by source information system.

[227] NIST, ibid.

[228] Codd, 1970, 1972, 1974; Date et al., 2003; Fagin, 1977, 1979, 1981 ibid.

**Assumption 2:** Experts in the practice of source information systems, not just the design, are on hand to be consulted on source information context.

> Should an information source context be unclear, experts are on hand to clarify matters and externalize the context in which source information carries meaning.

Nonaka[229]proposes a theory of organizational knowledge creation, which rests on three principles: creative chaos, redundancy, and requisite variety. Within this theory, built on classical and contemporary epistemological endeavors (notably Plato, Descartes, Locke, Hume and Kantas well as Husserl, Piaget, Polanyi, Ryle, and Hayek[230]), knowledge builds on previous knowledge incrementally and is of two epistemological natures, tacit and explicit.

Nonaka describes four transitions: tacit-to-tacit Socialization, tacit-to-explicit Externalization, explicit-to-explicit Combination, and explicit-to-tacit Internalization.

Nonaka also describes an ontological nature to knowledge, embedding it at individual, group, organizational and multi-organizational levels and promotes the idea that organizational knowledge creation develops spirally within these two dimensions, four transitions, and time.

---

229 Nonaka, I. 1994. A dynamic theory of organizational knowledge creation. *Organization Science*, 5(1): 14.

230 Specifically, Nonaka refers to Plato's *Theaetetus and Phaedo*, Descartes's *Discourse on Method*, Locke's *An Essay Concerning Human Understanding*, Hume's *An Enquiry Concerning Human Understanding*, and Kant's *Critique of Pure Reason* and to Husserl, E. G. 2010. *The Ideas of Phenomenology*: Springer, Piaget, J. 1974. *Recherches sur la Contradiction.* Paris: Presses Universitaires de France, Polanyi, M. 1966. *The Tacit Dimension.* London: Routledge & Kegan Paul, Ryle, G. 1949. *The Concept of Mind.* London: Huchinson, von Hayek, F. A. 1945. The Use of Knowledge in Society. *American Economic Review*, 35(4): 519-530..

Nonaka views both top executives and lower rank-and-file staff members as peripheral to the knowledge creation process, and elevates middle management to the status of centrifugal mainstay of organizational knowledge in what he deems the middle-up-down organization model, which he confronts with the top-down classical, i.e. bureaucratic, model and the bottom-up, i.e. creative but chaotic, model.

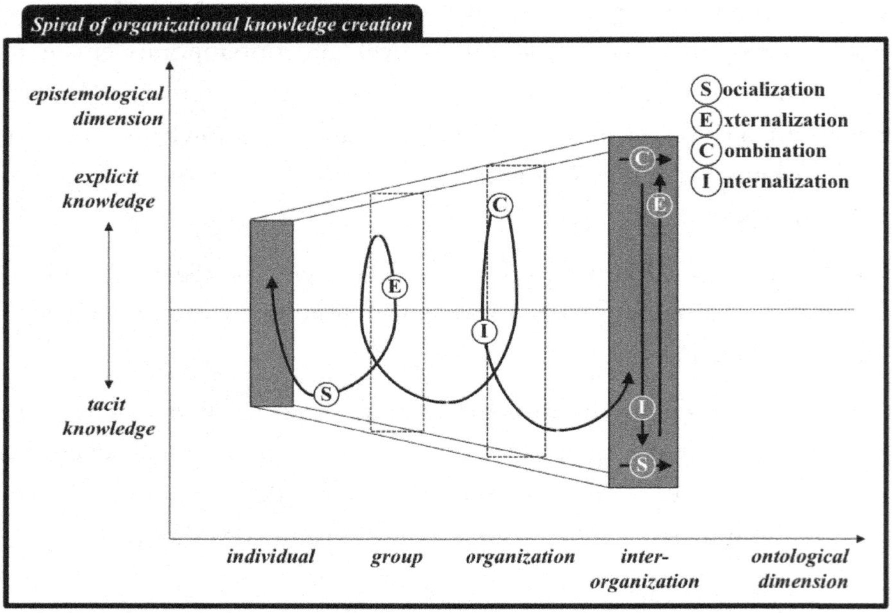


*Figure 26: Spiral of organizational knowledge creation*[231]

[231] Nonaka, I. 1994. A dynamic theory of organizational knowledge creation. *Organization Science*, 5(1): 14.

I suggest that information systems, as the visible part of organizational knowledge, may also be situated along the ontological dimension proposed by Nonaka. Specifically, we find (a) micro-information systems for the individual (occurring as Personal Information Managers or PIMs, or paper notebooks, calendars and address books…), (b) meso-information systems for teams and other groups (occurring as project wikis, departmental applications such as payroll software, or project collaborative software such as the now defunct GoogleWave…), and (c) macro-information systems for entire organizations or wider use (occurring as Enterprise Resource Planning packages or ERPs, corporate email systems, or market trading systems shared by Investment Banks).

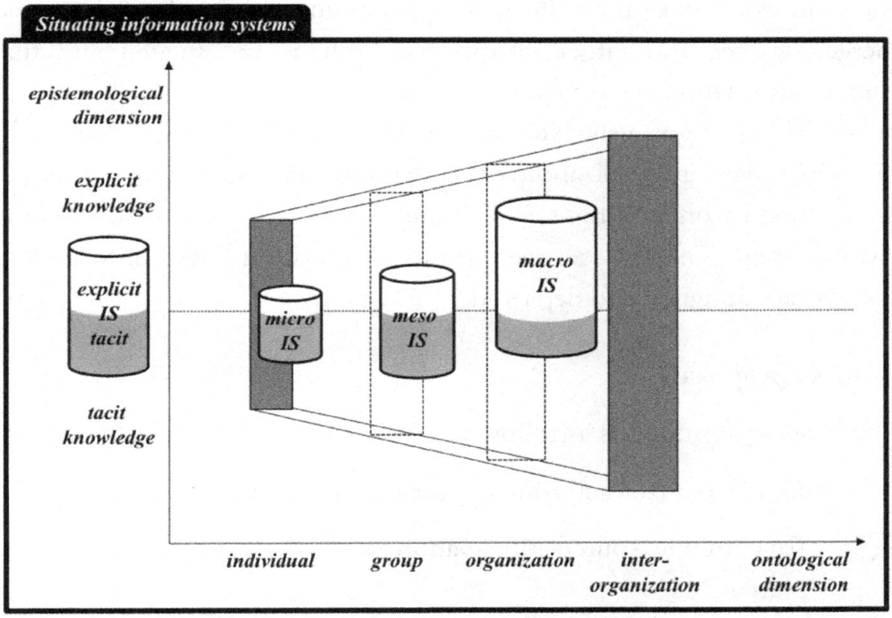


*Figure 27:Situating information systems epistemologically and ontologically*

Just as knowledge is dually tacit and explicit in varying measures, actuation of these information systems is enmeshed with their respective practitioners. A personal notebook, as an explicit manifestation of individual knowledge not meant to be shared (as opposed to a memo), may be difficult to interpret for another individual. A project wiki may be a bit of a cipher to someone outside the project team because of tacit convention. An ERP implementation may be hard for an organizational outsider to fathom unless implementation options are explained. A non-trader may be puzzled by what circulates on a trading system.

All in all, though, we may expect that the more macro an information system, the more of its knowledge is explicit; conversely, the more micro, the more tacit its knowledge. Figure 27 sums this situation up.

As assumed above, it is therefore paramount that, within a planned merger, source and target information systems be situated and that source information system experts on the practice and not just the theory be available to rigorously externalize each source information system. In practice today, internalization of target information systems is better understood by organizations, as it is implicitly involves implementing the change management exercises undertaken when new information systems are designed and deployed.

## Process proposals

The proposed process is as follows:

(a) externalize each source information system,

(b) combine source information systems by target information system,

(c) internalize the target information system to its new sponsor, the Emerging, Merging, Merged Organization.

To illustrate, the classic Suppliers and Parts database[232], is used (cf. Figure 28).

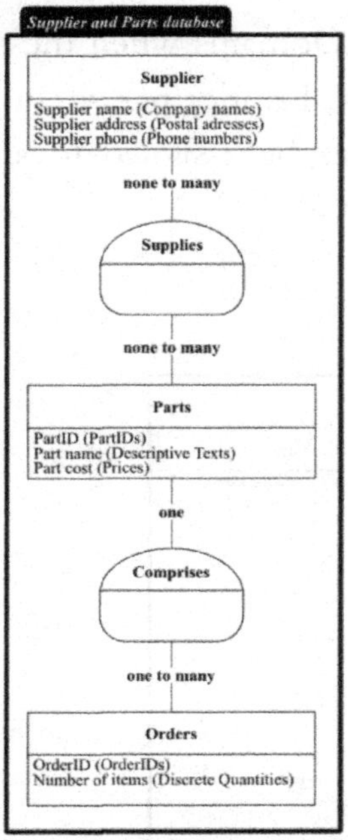


*Figure 28:Supplier and Parts database conceptual schema*

**Step 1: Externalizing each source information system**

Externalizing a source information system involves transforming its normalized conceptual schema through contextualization of every Attribute. This is the process for which I am coining the term '*archnormalization*'.

[232] Date, C. J. 2004. 8 Relational Calculus. In M. Suarez-Rivas, & K. Harutunian (Eds.), *An Introduction to Database Systems*: Pearson Education Inc.

**Archnormalizing Entities.** As exposed earlier, all Attributes have a social quality.

Attributes, assigned to Entities and Valued within Domains when these Entities are instantiated, are not properties of these Entities per se, but merely property a (social) Institution acknowledges be assigned to an Entity, to be valued according to a (social) Convention.

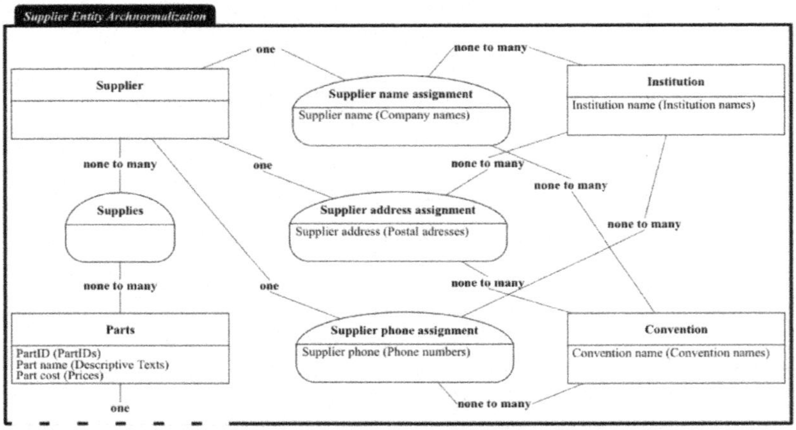


*Figure 29: Supplier and Parts database Supplier Entity Archnormalization*

In Figure 29, I make explicit the social assignment of Attributes to Suppliers by replacing each Supplier Entity Attribute (respectively Supplier Name, Address, and Phone) with the ternary Relationship that assigns each (respectively Supplier Name, Address, and Phone assignment).

In practice, Supplier Names may have been assigned by the Suppliers and Parts database custodian or by a consular authority, in accordance to an implicit or explicit naming norm. Until Supplier Names are shared or merged with other information systems, there is no need to make the convention explicit.

Similarly, Supplier Addresses and Phones will follow conventions and, at their most informal would consist respectively of freeform text in any language and freeform numbers, or, at their most formal would comply, on Earth, respectively with Universal Postal Union and International Telecommunication Union practices.

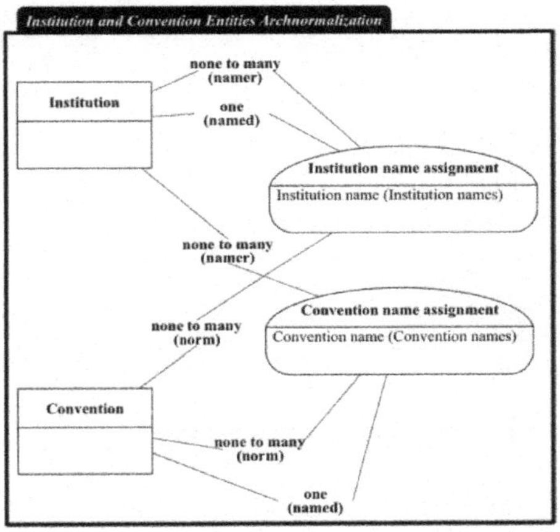


*Figure 30: Institution and Convention Entities Archnormalization*

Naturally, the Institutions and Conventions would themselves be archnormalized, as illustrated in Figure 30. Institutions name other Institutions, and themselves if they are sovereign. Conventions guide the naming of Institutions and of other Conventions and themselves. At the looser end of the spectrum, a naming convention may simply imply the use of a given alphabet.

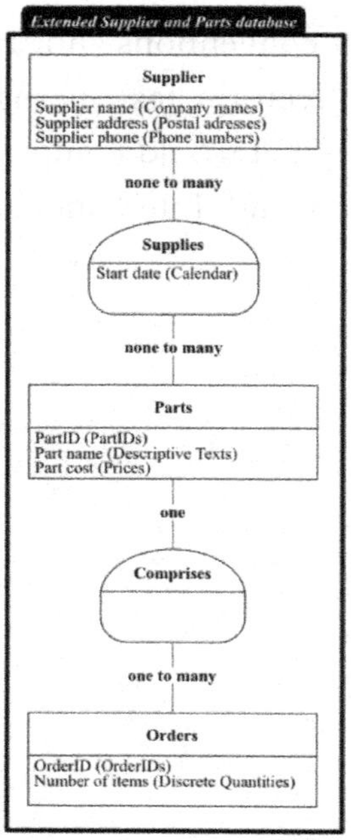


*Figure 31: Extended Supplier and Parts database (with a Relationship Attribute)*

Since no Relationship in the Suppliers and Parts database bears an Attribute, I've added a Start date in the '(Supplier) Supplies (Part)' Relationship to describe when a Supplier includes a Part in its catalog. The extended Suppliers and Parts database is illustrated in Figure 31.

**Archnormalizing Relationships.** Relationship Attributes, assigned to Relationship and also Valued within Domains when these Relationships are instantiated, are properties of these Relationships, but their expression requires an Institution acknowledges their assignment to these Relationships, the compliance of their valuation to a Convention. Figure 32 illustrates this process in my extended Suppliers and Parts database.

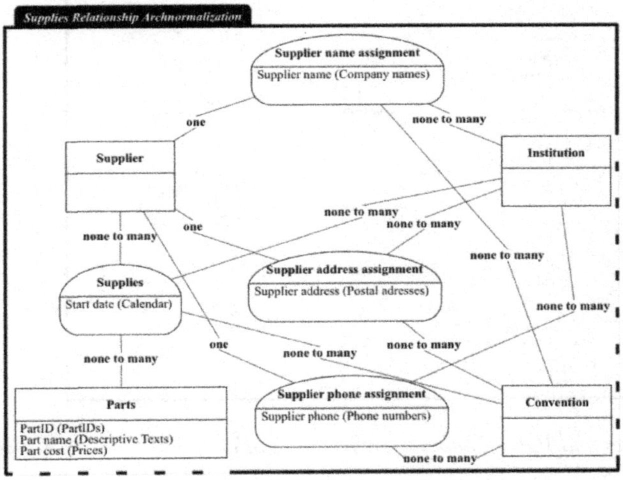


*Figure 32: SuppliesRelationship Archnormalization*

The process started with the Supplier Entity and the '(Supplier) Supplies (Part)' Relationship has to be generalized to all Entities and Relationships in the schema. Figure 33 presents the resulting archnormalized schema.

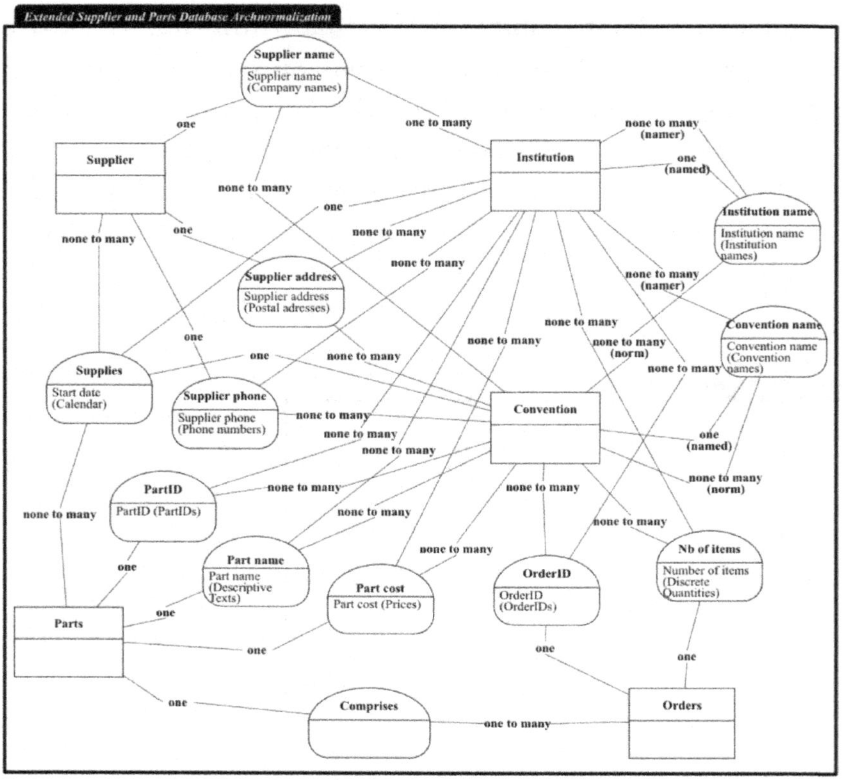


*Figure 33: Extended Supplier and Parts database Archnormalization*

## Midpoint Discussion

Scrutiny of this archnormalized schema illustrates that Entities, or concepts in laymen's terms, are inherently free of Attributes, and that, fundamentally, from a sociological perspective, each Entity is a nexus of potential Relationships. All concepts —all ideals[233] — are nexuses of potential connections with other concepts or, in the case of potential reflexive connections, the same concept. A streamlined conceptual model in archnormal form, i.e. with Attribute-free Entities, might be displayed as in Figure 34.

---

233 Jullien, F. 2009. *L'invention de l'idéal et le destin de l'Europe.* Paris: Seuil.

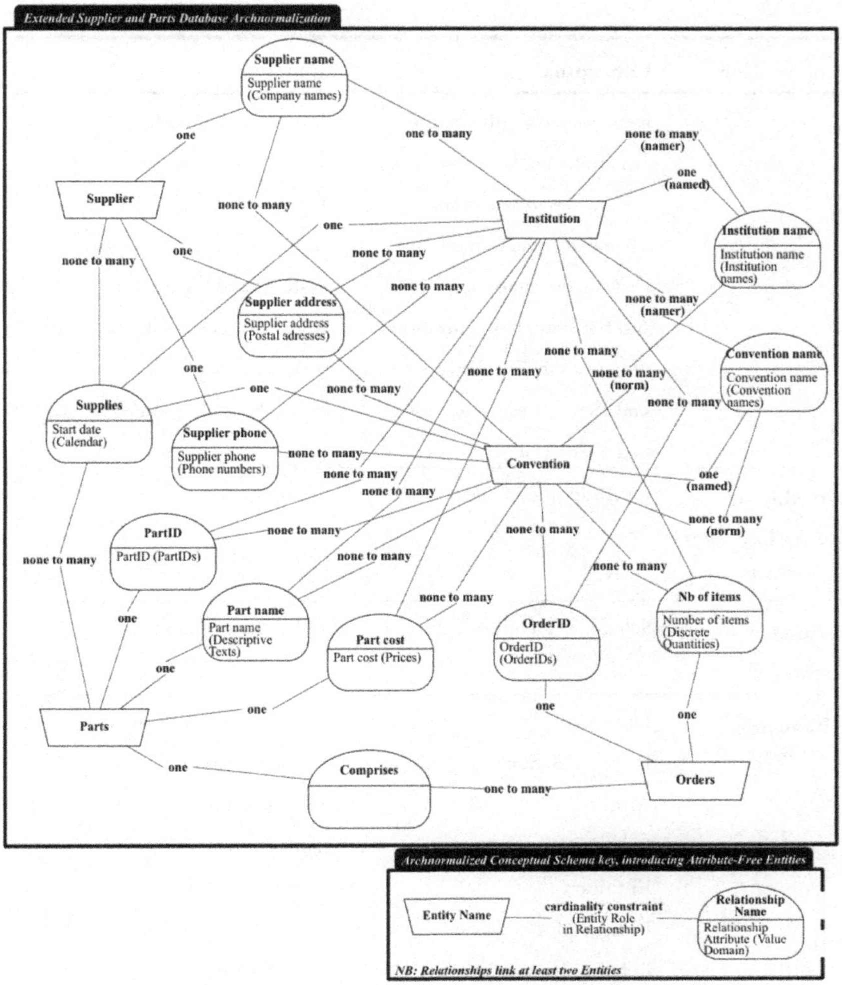


*Figure 34: Extended Supplier and Parts database Archnormalization, with Attribute-free Entities key*

| Level of Analysis | Conceptual | Concrete |
|---|---|---|
| **Scope** | Expression of Information System through a Conceptual schema, *neutral* from the *internal* organization of data custody by database systems and the *external* organization of collection and delivery to said IS users and other ISs said IS interfaces to. | Expression of particular Information System, *neutral* from the *internal* organization of data custody by database systems and the *external* organization of collection and delivery to said IS users and other ISs said IS interfaces to. |
| **Relationship to counterpart in other level** | Idealization of Object | Instantiation of Concept |
| **Conformance assumption** | Schema is archnormalized | IS has structural integrity |
| **Core Elements** | Domains<br>Domain hierarchies<br>Attribute-free Entities<br>Relationships between Entities<br>Entity Roles between Relationships and Related Entities<br>Cardinality Constraints between Relationships and Related Entities<br>Relationship Attributes<br>Attribute Domain-casting | Value Ranges (w/Units)<br>Range hierarchies<br>Tuples (Table rows)<br>Tuple Primary Keys (to identify tuples)<br>Tuple Foreign Keys (to identify links)<br>Tuple fields (Column values)<br>Field Types (Value ranges) |

*Table 11: Attribute-free Entities and Information-Less Objects analysis at conceptual and concrete levels*

Leaving the ideal plane for the object plane (or the conceptual schema for instantiations of its Entities, Relationships and their valued Attributes), objects are intrinsically devoid of information and nexuses of values-bearing links. Indeed, the organizational cognition behavior of object categorization, in which individuals and groups engage in, can be understood as a pattern recognition process enabling objects with comparable values-bearing links to emerge as a new category. Different groups observing, or paying attention to, different stimuli would, by virtue of repetitive acknowledgement, develop different categories and different vocabularies.

Table 11 summarizes and contrasts these musings on Attribute-free Entities at the ideal level of analysis, and Information-Less Objects from the object level of analysis.

**Archnormalizing Domains.** I now reprise the earlier Aristotelian breakdown of language into its component parts (cf. Table 1).

Aristotle[234] broke information attributes into four separate kinds: (a) enumerated information, (b) spatial information, (c) temporal information, and (d) coded information. Benveniste[235] suggests Ancient Greek's power of expression thus limited Aristotle's scope of thought.

Enumerated information is information representing a quantity of something being counted, the elements of which are indistinguishable from one another (i.e. *1*, *2*, or *31*).

A measure of length, geometric angle, weight, proportion, price, etc. is not enumerated information, as it implicitly suggests a comparison with a reference or standard. In Aristotle's view, such information is coded. Similarly, negative numbers (e.g. *-3*), percentages (e.g. *42%*), fractions (e.g. *¾*), algebraic expressions (e.g. $\sqrt{2}$) and geometric symbols (e.g. $\pi$) are objects of convention with no tangible reality: they are socially constructed abstractions. Spatial information (e.g. *here*, *above me*) conveys a physical reality. Temporal information (e.g. *now*, *recently*) conveys a temporal reality. This reality is always subjective, though its subjectivity may be collectively shared. Coded information is information coded according to a pre-set convention (e.g. *the number 0*, *felines*, *the DDR*, *organizational justice*, *in the Rockies*, *on Alpha Centauri*, *November 22nd 1963*, *Ithaca*, or *15 lbs.*); codes are the most common type of information and, as the existence of a pre-set convention implies, the most blatantly socially constructed of the four kinds of information attributes earmarked by Aristotle.

---

234 Aristotle. 2000a. *Catégories de l'interprétation: Organon I et II*: Vrin.

235 Benveniste, E. 1966. *Problèmes de linguistique générale*. Paris: Gallimard.

Information is sometimes implicitly carried. Use of the future tense is an example of implicitly carried temporal information. Enumerations represent the only absolutely defined category. Spatial, temporal information share with enumerations their universal meaning; however, as they express space or time relative to the speaker, they do not carry an absolute meaning. All others forms of information require a shared context, beyond a common language, to be understood.

I've demonstrated here above the characterization of (a) enumerations as universal and absolute, (b) space as universal and relative, (c) time as universal and relative, and (d) all other language components to be socially constructed, as of a Western, or more precisely since language is the level of analysis, of an Indo-European idiosyncratic nature.

Therefore, a Domain, regardless of whether it is expressed numerically, spatially, or temporally, warrants an explicit institutional frame of reference.

Table 12 sums up a systematic method to generate default, explicit domains, though externalizing the tacitly practiced domains will help an information system merger's success.

| | Code, locally shared | Code, universally accepted |
|---|---|---|
| **Top Domains** | e.g. customer ID, supplier status… | e.g. mutually agreed barcode; media; postal address; phone number; price; Julian, Japanese, or Muslim date; color; weight; discrete or continuous quantity |
| **Default Institution** | Information system organizational sponsor e.g. ACME accounts payable dept, ACME Central procurement… | National, International, or Trade Standards Body e.g. Catastrophe Risk Evaluating and Standardizing Target Accumulations, *Institut National de la Propriété Industrielle*, International Standards Organization, International Telecommunication Union, Universal Postal Union, World-Wide Web Consortium |
| **Default Convention** | Information system organizational custodian coding convention e.g. ACME customer ID scheme, ACME 3rd party status taxonomy… | Standard coding convention e.g. CRESTA zoning, JPEG, ISO-434, European Patent Numbering, Simplified Chinese business jargon, XBRL, XML |

*Table 12: Domain Archnormalization alternative strategies*[236]

[236] In this table, a hypothetical organization ACME is envisaged.

## Step 2: Combining archnormalized information systems

**Combining Domains.** At this stage, Domains originating from source information systems should be merged by meshing the original hierarchies into target hierarchies.

If top source domains semantically overlap, arbitration is necessary. Simple, common sense rules should be followed, such as:

(a) opt for globally-shared over locally-shared domains wherever possible (use of chassis number for vehicle should be preferred over a company specific vehicle identification scheme);

(b) if concurrent, globally-shared domains emerge (e.g. tons, kgs, and lbs, or ISO 3166-1 alpha-2 and ISO 3166-1 alpha-3…), have the target information system custodian's arbitration body rule on which one to retain,

(c) if locally-shared domains overlap semantically, study the three following options:

(i) study whether a globally-shared domain has emerged since these domains' local inception in their source information system;

(ii) retain the domain with the highest resolution if all other domains map unambiguously to it;

(iii) combine the domains into a supra-domain that all source domains map to.

**Combining Entities.** As this stage, the union (in the set algebra sense of the word) of Entities originating from source information systems will represent Entities in the target information system. If a particular Entity is present in several source information systems, different primary keys may have been used by different information systems to identify Entities. A choice of primary key must be made, and the redundant primary keys become alternate keys.

Figure 35 illustrates of the reconciliation of duplicate entity instances.

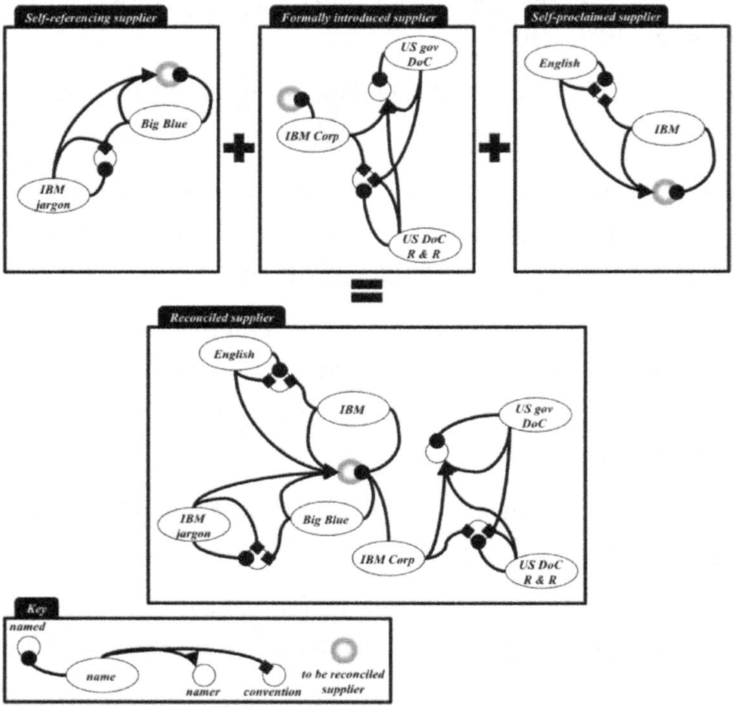


*Figure 35: An example Entity instance (a Supplier variously known as IBM, IBM Corp and Big Blue) being addressed by different names*

This general rule applies of course to the newly minted Entities that are Institution and Convention, though overlap may be slight in the first archnormalizations.

**Combining Relationships.** As this stage, the union of Relationships originating from source information systems will represent Relationships in the target information system. If a particular Relationship is present in several source information systems, different foreign keys may have been used by different information systems to identify Entities. The new choices of Entity primary keys must be systematically enforced in Relationships.

**Combining Relationship Attributes.** As this stage, the merging of Relationships may have resulted in aggregating Attributes from several source information systems within the same target Relationships. Some of these Attributes may overlap: the characteristics of the Attributes' respective source Domains and how these Domains were combined in the target information system should guide the overlapping Attribute combination.

### Step 3: Internalizing each target information system

Internalizing each target information system is a classical change management exercise. There need not be any further expansion within this paper of this already well-explored topic[237].

[237] Interested readers are invited to peruse for example Spector, B. A. 2006. *Implementing Organizational Change: Theory Into Practice*: Prentice-Hall..

## Conclusion: Remember the Analogue!

As Nonaka[238] reminds us, Bateson[239] has referred to tacit knowledge gathering as the *analogue* parallel processing of the mind, the counterpart to the *digital* formalization of explicit knowledge that sequential computing embodies. Given its digital, codified nature, formalizable knowledge is countably infinite. Conversely, as noted in the previous chapter[240], the analogue range of potential knowledge we aspire to grasp and fathom is uncountably infinite, making it of a higher order of infinity.

Organizations should therefore ponder this humbling thought and always remember to balance their digital endeavors with the analogue encompassment key to a venture's success.

---

[238] Nonaka, I. 1994. A dynamic theory of organizational knowledge creation. *Organization Science*, 5(1): 14.

[239] Bateson, G. 2000. *Steps to an Ecology of Mind.* Chicago: University of Chicago Press.

[240] Leveraging once more the diagonal argument of Cantor, G. 1874. Ueber eine Eigenschaft des Inbegriffes aller reellen algebraischen Zahlen. *Journal für die reine und angewandte Mathematik*, 77: 258-262.

# X] FUTURE DIRECTIONS

Sections of this thesis have already delved into three implications of this thesis' central findings:

- the devising of methodological recommendations for building math tools from social phenomena, rather than using the math social scientists, including economists and business modellers, know, in a fashion already embraced by theoretical physicists, including the operationalization of social phenomena with a toolset adapted from quantum physicists (see sections V and VI); work here is far from over;
- implications for qualitative research methods and notable discourse analysis (see section VII); modelling not only what concepts support narratives but their cultural and disciplinary origination and how these are situated and interplay in the cognitive schemas of the narrator; the introduced methodology has been peer-reviewed and endorsed for trials by several scholars; however, both the robustness and effectiveness of the proposed methodology have to be established;
- implications for information systems mergers (see section IX), which have been prototyped, but for which no extensive field tests have been conducted at this stage.

Further research currently sketched out includes:

- development of an extension to an established[241] computer model of learning, that introduces semiotic heterogeneity in learning; this will simulate exploration, exploitation in cross-cultural, cross-disciplinary social networks, on which work is collaboratively ongoing in parallel to this thesis;

---

[241] Lazer, D., & Friedman, A. 2007. The Network Structure of Exploration and Exploitation. *Administrative Science Quarterly*, 52(4): 667-693. based on genetic algorithms

- an empirical measure of professional jargon distance based on corporate parameterizations of machine translation and enterprise search engines, in the contexts of organic growth and of mergers, for which empirical analysis is currently pending legal approval.

---

and rugged landscapes Levinthal, D. A. 1997. Adaptation on rugged landscapes. *Management Science*, 43(7): 934-950.

# AFTERWORD

The conclusion brought forth hereabove assimilating all formal knowledge to codes, which in effect, may be enumerated and thereby reduced to 1s and 0s mustn't blind us to the fact that sentient beings such as ourselves also experience beauty in all we perceive. Langer[242], in her seminal opus, proposed that (a) not only is language the form of symbols of things; (b) language also is the tool through which humans can think; (c) (primeval) thought precedes languages and is the seat of aesthetic appreciation. Ramirez[243] demonstrated how people, armed with this very sense of aesthetics, even experience beauty (and, symmetrically, ugliness) in such immaterial forms as organizations; indeed, Ramirez observed repeatedly that successful social organizations, implicitly or not, cultivated a form persistently perceived as beautiful by the volunteers and patrons that kept committing their time and resources to the perpetuation of that most fluid of organizations[244]. That some instinctively recoil at the thought of contributing to some collective form of action (be it a social organization, a for-profit firm, or a sporting event), while others do not, illustrates the diverse sense of aesthetics to be found amongst us.

Beauty precedes formulated thought. Beauty may be found everywhere and appreciation of aesthetic form cannot be reduced to code, i.e. to 1s and 0s.

---

242 For more on this, cf.

Langer, S. K. K. 1948. *Philosophy in a new key, a study in the symbolism of reason, rite and art* ([1st ed.). New York,: Penguin Books.

Langer, S. K. K. 1953. *Feeling and form; a theory of art.* New York,: Scribner.

243 Ramírez, R. 1991. *The beauty of social organizations.* Munich: Accedo Verlagsgesellschaft.

244 Indeed I may argue later that volunteer or social organizations are where social scientists may observe human interactions at its purest, least hindered by legal distortions (or at least labor law distortions).

# Appendixes

## Social Network Theory Definitions

*Balance theory* is used to predict the existence of closed triads (e.g. A--B, B--C, and A--C) at greater-than-chance levels in human networks. It is an important premise of Granovetter's strength of weak ties argument.

*Centrality* refers to a family of structural concepts relating to a node's position in the network.

Cohesion is generally thought of as a group-level property, reflecting the connectedness of the group network. However, it is convenient to also think of *dyadic-level cohesion*, which reflects the strength or closeness of a pair of nodes. In older network literature, dyadic cohesion referred to as "relational cohesion" or as "social proximity".

Experimental *exchange* networks is an area of study in which subjects are brought into a lab and asked to play a game in which they must try to get as many points as possible. In each round, they are given 24 points to divide up with an exchange partner. The experimenter arranges the people into networks of who can exchange with whom. Initially, it was thought that centrality would determine point-getting, but this is not true.

*Group cohesion* and shape are both group-level (i.e. network-level) concepts. Group cohesion refers to the overall connectedness or knittedness of a network. Shape measures indicate what kind of shape the network has, such as core-periphery, clumpy, etc.

*Homophily* refers to the tendency for people to have (non-negative) ties with people who are similar to themselves in socially significant ways. The term itself (due to Lazarsfeld) specifically refers to an internal preference. However, today it is usually used to refer to an observable behavioral tendency whose causes can include preference, but can also include opportunity (e.g. you have many white friends because most people around you are white).

*Cohesive regions* and *subgroups* refer to a family of theoretical concepts relating to the sociological concept of group.

*Regular equivalence* is an attempt to capture the sociological notion of a relational or structural role. Two nodes are said to play the same role (i.e. are regularly equivalent) if the have ties to the same roles. For example, two nurses will have similar types of relations (as each other) to doctors, patients, other nurses, and so on. Similarly, two patients tend to have similar relations as each other with doctors, nurses, etc. Important note: the equivalent actors has the same types of relations with equivalent others, but not necessarily the same others.

*Social capital* refers to the benefits of social ties or position in a social network. The concept may be applied at both the actor and group levels.

Loosely speaking, *structural equivalence* refers to the extent to which two nodes are connected to the same others —i.e. have the same social environments. It is often hypothesized that structurally equivalent nodes will be similar in other ways as well, such as in attitudes, behaviors or performance.

## Information modelling definitions[245]

Alias: A nonstandard name for an entity or domain (attribute).

Assertion: A statement that specifies a condition that must be true.

Attribute: A property or characteristic that is common to some or all of the instances of an entity. An attribute represents the use of a domain in the context of an entity.

Attribute Value: A value given to an attribute in an entity instance.

Attribute, Inherited: An attribute that is a characteristic of a category entity by virtue of being an attribute in its generic entity or a generic ancestor entity.

Attribute, Migrated: A foreign key attribute of a child entity.

Attribute, Non-key: An attribute that is not the primary or a part of a composite primary key of an entity. A non-key attribute may be a foreign key or alternate key attribute.

Attribute, Optional: A non-key attribute of an entity that may be null in any instance of the entity.

Attribute, Owned: An attribute of an entity that has not migrated into the entity.

Category Cluster: A set of one or more mutually exclusive categorization relationships for the same generic entity.

Category Discriminator: An attribute in the generic entity (or a generic ancestor entity) of a category cluster. The values of the discriminator indicate which category entity in the category cluster contains a specific instance of the generic entity. All instances of the generic entity with the same discriminator value are instances of the same category entity. The inverse is also true.

---

245 (National Institute of Standards and Technology, 1993).

Conceptual Schema: See Schema

Constraint: A rule that specifies a valid condition of data.

Constraint, Cardinality: A limit on the number of entity instances that can be associated with each other in a relationship.

Constraint, Existence: A condition where an instance of one entity cannot exist unless an instance of another related entity also exists.

Database: A collection of interrelated data, often with controlled redundancy, organized according to a schema to serve one or more applications.

Data Model: A graphical and textual representation of analysis that identifies the data needed by an organization to achieve its mission, functions, goals, objectives, and strategies and to manage and rate the organization. A data model identifies the entities, domains (attributes), and relationships (or associations) with other data, and provides the conceptual view of the data and the relationships among data.

Data Type: A categorization of an abstract set of possible values, characteristics, and set of operations for an attribute. Integers, real numbers, character strings, and enumerations are examples of data types.

Domain: A named set of data values (fixed, or possibly infinite in number) all of the same data type, upon which the actual value for an attribute instance is drawn. Every attribute must be defined on exactly one underlying domain. Multiple attributes may be based on the same underlying domain.

Enterprise: An organization that exists to perform a specific mission and achieve associated goals and objectives.

Entity: The representation of a set of real or abstract things (people, objects, places, events, ideas, combination of things, etc.) that are recognized as the same type because they share the same characteristics and can participate in the same relationships.

Entity, Category: An entity whose instances represent a sub-type or sub-classification of another entity (generic entity). Also known as sub-type or sub-class.

Entity, Child: The entity in a specific connection relationship whose instances can be related to zero or one instance of the other entity (parent entity).

Entity, Generic: An entity whose instances are classified into one or more sub-types or sub-classifications (category entity). Also known as super-type or super-class.

Entity Instance: One of a set of real or abstract things represented by an entity. The instance of an entity can be specifically identified by the value of the attribute(s) participating in its primary key.

Entity, Parent: An entity in a specific connection relationship whose instances can be related to a number of instances of another entity (child entity).

Existence Dependency: A constraint between two related entities indicating that no instance of one (child entity) can exist without being related to an instance of the other (parent entity). The following relationship types represent existence dependencies: identifying relationships, categorization relationships and mandatory non-identifying relationships.

External Schema: See Schema

Functional Dependency: A special kind of integrity constraint that applies within the confines of a single entity "R", where each "X" value of "R" has associated with it at most one "Y" value of "R" (at any one time). Attributes "X" and "Y" may be composite.

Glossary: A set of definitions of entities and domains (attributes).

IDEF1X Diagram: See View Diagram

IDEF1X Model: A set of one or more IDEF1X views, often represented as view diagrams, which depict the underlying semantics of the views, along with definitions of the entities and attributes used in the views. See Data Model.

Identifier Dependency: A constraint between two related entities that requires the primary key in one (child entity) to contain the entire primary key of the other (parent entity). The following relationship types represent identifier dependencies: Identifying relationships, categoryzation relationships.

Key, Candidate: An attribute, or combination of attributes, of an entity whose values uniquely identify each entity instance.

Key, Alternate: Any candidate key of an entity other than the primary key.

Key, Composite: A key comprised of two or more attributes.

Key, Compound: Same as Key, Composite

Key, Foreign: An attribute, or combination of attributes of a child or category entity instance whose values match those in the primary key of a related parent or generic entity instance. A foreign key results from the migration of the parent or generic entities primary key through a specific connection or categorization relationship.

Key, Migrated: Same as Foreign Key

Key Migration: The modeling process of placing the primary key of a parent or generic entity in its child or category entity as a foreign key.

Key, Primary: The candidate key selected as the unique identifier of an entity.

Key, Split: A foreign key containing two or more attributes, where at least one of the attributes is a part of the entities primary key and at least one of the attributes is not a part of the primary key.

Normal Form: The condition of an entity relative to satisfaction of a set of normalization theory constraints on its attribution. A specific normal form is achieved by successive reduction of an entity from its existing condition to some more desirable form. The procedure is reversible.

- First Normal Form (1NF) - An entity is in 1NF if and only if all underlying simple domains contain atomic values only.
- Second Normal Form (2NF) - An entity is in 2NF if and only if it is in 1NF and every non-key attribute is fully dependent on the primary key.
- Third Normal Form (3NF) - An entity is in 3NF if and only if it is in 2NF and every attribute that is not a part of the primary key is non-transitively dependent (mutually independent) on the primary key. Two or more attributes are mutually independent if none of them is functionally dependent on any combination of the others.

Normalization: The process of refining and regrouping attributes in entities according to the normal forms.

Null: A condition where a value of an attribute is not applicable or not known for an entity instance.

Relationship: An association between two entities or between instances of the same entity.

Relationship Cardinality: The number of entity instances that can be associated with each other in a relationship. See Constraint, Cardinality.

Relationship, Categorization (Category): A relationship in which instances of both entities represent the same real or abstract thing. One entity (generic entity) represents the complete set of things the other (category entity) represents a sub-type or sub-classification of those things. The category entity may have one or more characteristics, or a relationship with instances of another entity not shared by all generic entity instances. Each instance of the category entity is simultaneously an instance of the generic entity.

Relationship, Connection: Same as Relationship, Specific Connection

Relationship, Identifying: A specific connection relationship in which every attribute in the primary key of the parent entity is contained in the primary key of the child entity.

Relationship, Mandatory Non-identifying: A non-identifying relationship in which an instance of the child entity must be related to an instance of the parent entity.

Relationship Name: A verb or verb phrase which reflects the meaning of the relationship expressed between the two entities shown on the diagram on which the name appears.

Relationship, Non-specific: A relationship in which an instance of either entity can be related to a number of instances of the other.

Relationship, Non-identifying: A specific connection relationship in which some or all of the attributes contained in the primary key of the parent entity do not participate in the primary key of the child entity.

Relationship, Optional Non-identifying: A non-identifying relationship in which an instance of the child entity can exist without being related to an instance of the parent entity.

Relationship, Parent-Child: Same as Relationship, Specific Connection

Relationship, Specific Connection: A relationship where a number of instances of one entity (child entity) can be related to zero or one instance of the other entity (parent entity). In a specific connection relationship the primary key of the parent entity is contributed as a foreign key to the child entity.

Role Name: A name assigned to a foreign key attribute to represent the use of the foreign key in the entity.

Schema: A definition of data structure:

- Conceptual Schema: A schema of the ANSI/SPARC Three Schema Architecture, in which the structure of data is represented in a form independent of any physical storage or external presentation format.
- External Schema: A schema of the ANSI/SPARC Three Schema Architecture, in which views of information are represented in a form convenient for the users of information; a description of the structure of data as seen by the user of a system.
- Internal Schema: A schema of the ANSI/SPARC Three Schema Architecture, in which views of information are represented in a form specific to the data base management system used to store the information: a description of the physical structure of data.

Semantics: The meaning of the syntactic components of a language.

Synonym: A word, expression, or symbol accepted as a figurative or symbolic substitute for another word or expression; that is, an alternative name for the same thing. (See Alias)

Syntax: Structural components or features of a language and rules that define relationships among them.

Verb Phrase: A phrase used to name a relationship, which consists of a verb and words which comprise the object of the phrase.

View: A collection of entities and assigned attributes (domains) assembled for some purpose.

View Diagram: A graphic representation of the underlying semantics of a view.

## Database normalization further definitions[246]

An *attribute* is as property or characteristic common to all instances of a relation. An attribute represents the use of a domain in the context of a relation.

> In normalized relational databases, attributes are instantiated as table columns and represent either recorded entity properties (e.g. a Person's Gender), or recorded relationship properties (e.g. an attribute of the Company Employs Person relationship could be the Inception and Expiry dates of the employment).

A *foreign key* is an attribute, or combination of attributes of a child or category entity instance whose values match those in the primary key of a related parent or generic entity instance. A foreign key results from the migration of the parent or generic entities primary key through a specific connection or categorization relationship.

> In normalized relational databases, primary keys are instantiated as table columns.

A *primary key* within an entity is the candidate key selected as the unique identifier of that entity.

> In normalized relational databases, primary keys are instantiated as table columns.

A *relation* is a set of tuples sharing attributes.

> In normalized relational databases, relations are instantiated as tables and represent either acknowledged entities (e.g. Persons, Companies, Products, Locations, Projects…), or acknowledged relationships (e.g. Company Employs Person, Company Carries Products, Person Resides at Location, Person Oversees Person within Project…).

---

[246] Synthesized from (Codd, 1970, 1972, 1974; Date et al., 2003; Fagin, 1977, 1979, 1981).

A *tuple* is an element of a relation.

In normalized relational databases, tuples are instantiated as table rows and represent either recorded entities (e.g. the Person recorded in the database whose primary key is #1234) or recorded relationships (e.g. that the Company recorded in the database whose primary key is #abcd employs the Person recorded in the database whose primary key is #1234 would be recorded in an Employs tuple).

# Bibliography


Abrahamson, E. 1996. Management fashion, academic fashion, and enduring truths. *Academy of Management Review*, 21(3): 616.

Adler, N. J. 1983. A Typology of Management Studies Involving Culture. *Journal of International Business Studies*, 14(2): 29-47.

Akerlof, G. A. 1970. The Market for "Lemons": Quality Uncertainty and the Market Mechanism. *The Quarterly Journal of Economics*, 84(3): 488-500.

Anderson, J. A., & Bell, J. M. 1997. *Number Theory With Applications*: Prentice Hall.

anon. 1879. *诗经 aka 詩經 (Shi Jing) aka The Book of Odes, Le Livre des Odes* (J. Legge, & M. Granet, Trans.).

anon. 1967. *易经 aka 易經 (Yi Jing), aka I Ching, the Book of Changes* (R. Wilhelm, C. F. Baynes, & E. Perrot, Trans.).

Aoki, M. 2004. An organizational architecture of T-form: Silicon Valley clustering and its institutional coherence. *Industrial and Corporate Change*, 13(6): 967.

Aristotle. 2000a. *Catégories de l'interprétation: Organon I et II*: Vrin.

Aristotle. 2000b. *Métaphysique* (J. Tricot, Trans.). Paris: Vrin.

Bateson, G. 2000. *Steps to an Ecology of Mind*. Chicago: University of Chicago Press.

Benveniste, E. 1966. *Problèmes de linguistique générale*. Paris: Gallimard.

Berger, P. L., & Luckmann, T. 1966a. *The social construction of reality; a treatise in the sociology of knowledge* ([1st ed.). Garden City, N.Y.,: Doubleday.

Berger, P. L., & Luckmann, T. 1966b. *The Social Construction of Reality: A Treatise in the Sociology of Knowledge*: Anchor Books.

Bernard, H. R. 2006. Honoring Peter Killworth's contribution to social network theory. University of Southampton.

Bernard, H. R., Shelley, G. A., & Killworth, P. D. 1987. How much of a network does the GSS and RSW dredge up? *Social Networks*(9): 49-61.

Berry, J. W., Poortinga, Y. H., Segall, M. H., & Dasen, P. R. 2002. *Cross-cultural psychology: research and applications* (2nd ed.): Cambridge University Press.

Bitbol, M. (Ed.). 2009. *Théorie quantique et sciences humaines*. Paris: CNRS.

Bitbol, M. 2010. Théorie quantique et sciences humaines — Les savoirs du non-séparé.

Bourguignon, J.-P. 2010. Espaces courbes de Gauss à Perelman, en passant par Einstein. In S. M. d. France (Ed.), *Un texte, un mathématicien*.

Brown, J. S., & Duguid, P. 1991. Organizational Learning and Communities-of-Practice: Toward a Unified View of Working, Learning, and Innovation. *Organization Science*, 2(1): 40-57.

Butts, C. T. 2008a. network: a Package for Managing Relational Data in R. *Journal of Statistical Software*, 24(2).

Butts, C. T. 2008b. Social Network Analysis with sna. *Journal of Statistical Software*, 24(6).

Butts, C. T. 2009. sna: Tools for Social Network Analysis. R package version 2.0-1.

Butts, C. T., Handcock, M. S., & Hunter, D. R. 2008. network: Classes for Relational Data. R package version 1.4-1. Irvine, CA.

Cantor, G. 1874. Ueber eine Eigenschaft des Inbegriffes aller reellen algebraischen Zahlen. *Journal für die reine und angewandte Mathematik*, 77: 258-262.

Chabrol, C., Courbet, D., & Fourquet-Courbet, M.-P. 2004. Psychologie sociale, traitements et effets des médias, Vol. 2008.

Codd, E. F. 1970. A Relational Model of Data for Large Shared Data Banks. *Communications of the ACM*, 13(6): 377–387.

Codd, E. F. 1972. Further Normalization of the Data Base Relational Model. In R. J. Rustin (Ed.), *Data Base Systems: Courant Computer Science Symposia Series 6*: Prentice-Hall.

Codd, E. F. 1974. Recent Investigations into Relational Data Base Systems (IBM Research Report RJ1385), *Proc. 1974 Congress*. New York: North-Holland.

Cohen, M. D., March, J. G., & Olsen, J. P. 1972. A Garbage Can Model of Organizational Choice. *Administrative Science Quarterly*, 17(1): 1-25.

Confucius. 2010. *论语 aka 論語 (Lun Yu) aka The Analects* (S. Couvreur, D. C. Lau, & J. Legge, Trans.).

Cook, R. T. 2006. Knights, knaves and unknowable truths. *Analysis*, 66(289): 10-16.

Copernicus, N. 1543. *De revolutionibus orbium coelestium.*

Cowen, R. S. 1985. How the Refrigerator Got Its Hum. In D. MacKenzie, & J. Wajcman (Eds.), *The Social Shaping of Technology*: 208-211. Buckingham: Open University Press.

Crosby, A. W. 1997. *The Measure of Reality: Quantification and Western Society, 1250-1600*: Cambridge University Press.

Cruse, A. D. 2004. *Meaning in Language.* Oxford: Oxford University Press.

Cullen, C. 2004. *The Suàn shù shu 筭數書 'Writings on reckoning': a translation of a Chinese mathematical collection of the second century BC, with explanatory commentary.* Cambridge: Needham Research Institute.

Date, C. J. 2004. 8 Relational Calculus. In M. Suarez-Rivas, & K. Harutunian (Eds.), *An Introduction to Database Systems*: Pearson Education Inc.

Date, C. J., Darwen, H., & Lorentzos, N. A. 2003. *Temporal Data and the Relational Model: A Detailed Investigation into the Application of Interval and Relation Theory to the Problem of Temporal Database Management.* Oxford: Elsevier.

Dawkins, R. 1989a. The Replicators, *The Selfish Gene.* Oxford: Oxford University Press.

Dawkins, R. 1989b. *The Selfish Gene.* Oxford: Oxford University Press.

Dedekind, R. 1888. *Was sind und was sollen die Zahlen? (What are and what should the numbers be?)*: Braunschweig.

Dehornoy, P. 2009. Cantor et les infinis. In S. M. d. France (Ed.), *Un texte, un mathématicien.*

Demazière, D., & Dubar, C. 1997. *Analyser les entretiens biographiques. L'exemple des récits d'insertion.* . Paris: Nathan.

DiMaggio, P. J., & Powell, W. W. 1983. The Iron Cage Revisited: Institutional Isomorphism and Collective Rationality in Organizational Fields. *American Sociological Review*, 48(2): 147-160.

Donaldson, L. 2010. *The Meta-Analytic Organization, introducing Statistico-Organizational Theory.* Armonk, NY: M.E Sharpe.

Dunbar, R. I. M. 1993. Coevolution of neocortical size, group size and language in humans. *Behavioral and Brain Sciences*, 16(4): 681-735.

Erdős, P., & Rényi, A. 1959. On Random Graphs. *Publicationes Mathematicae Debrecen*, 6: 290-297.

Espeland, W. N., & Stevens, M. L. 1998. Commensuration as a Social Process. *Annual Review of Sociology*, 24: 313-343.

Euclid. 1956. *The thirteen books of the elements* (T. L. Heath, Trans.): Dover Publications.

Euler, L. 1741. Solutio problematis ad geometriam situs pertinentis. *Commentarii academiae scientiarum imperialis Petropolitanae*(8): 128-140.

Euler, L. 1746. De seriebus divergentibus. *Opera Omnia*, I(14): 585-617.

Ewick, P., & Silbey, S. S. 1995. Subversive Stories and Hegemonic Tales: Toward a Sociology of Narrative. *Law & Society Review*, 29(2): 197.

Fagin, R. 1977. Multivalued Dependencies and a New Normal Form for Relational Databases. *ACM Transactions on Database Systems*, 2(1): 267.

Fagin, R. 1979. Normal Forms and Relational Database Operators, *International Conference on Management of Data*. Boston: ACM SIGMOD.

Fagin, R. 1981. A Normal Form for Relational Databases That Is Based on Domains and Keys. *Communications of the ACM*, 6(3): 387-415.

Fang, C., Lee, J., & Schilling, M. A. 2010. Balancing Exploration and Exploitation Through Structural Design: The Isolation of Subgroups and Organizational Learning. *Organization Science*, 21(3): 625-642.

Fendt, J. 2006. *The CEO in Post-Merger Situations: An Emerging Theory on the Management of Multiple Realities*. Leiden University, Leiden.

Fligstein, N. 1996. Markets as politics: A political-cultural approach to market institutions. *American Sociological Review*, 61(4): 656.

Fraser, C. 2009. School of Names. In E. N. Zalta (Ed.), *The Stanford Encyclopedia of Philosophy*, Summer 2009 ed.

Goffman, E. 1959. *The Presentation of Self in Everyday Life*: Anchor Books.

Granet, M. 1934. *La pensée chinoise*. Paris: Albin Michel.

Granovetter, M. 2007. The Social Construction of Corruption. In V. Nee, & R. Swedberg (Eds.), *On Capitalism*: 152-172: Stanford University Press.

Granovetter, M., & McGuire, P. 1998. The Making of an Industry: Electricity in the United States. In M. Callon (Ed.), *The Laws of The Markets*: 147-173. Oxford: Blackwell.

Gray, R. D., & Atkinson, Q. D. 2003. Language-tree divergence times support the Anatolian theory of Indo-European origin. *Nature*, 426: 435-439.

Harvard Law. 2003. RSS.

Hatch, M. J. 2006. *Organization Theory: Modern, Symbolic, and Postmodern Perspectives* (2 ed.): Oxford University Press USA.

Hedlund, G. 1994. A model of knowledge management and the N-form corporation. *Strategic Management Journal*, 15: 73.

Heisenberg, W. 1958. *Physics and Philosophy*: George Allen and Unwin Edition.

Hofstede, G. 1983. National Cultures in Four Dimensions: A Research-Based Theory of Cultural Differences Among Nations. *International Studies of Management and Organization*, 13(1,2): 46.

House, R. J., Hanges, P. J., Javidan, M., Dorfman, P., & Gupta, V. 2006. *Culture, leadership, and organizations: the GLOBE study of 62 societies*. Thousand Oaks, CA: Sage.

Husserl, E. G. 2010. *The Ideas of Phenomenology*: Springer.

Idelson, M., & Mangin, N. 2010. Learning in social networks: simulating the impact of different cognitive frames and interpretation schemes on exploration and exploitation. In S. A. Rodan (Ed.), *2010 Academy of Management Simulation in Organization Theory Professional Development Workshop*. Montréal.

Idelson, M., & Micaëlli, I. 2010. In the "I" of the innovative entrepreneur: a proposed qualitative analysis. In M. Graham, & Y. Li (Eds.), *CMS Organization workshop: Innovation and entrepreneurship track*. Montréal.

IMC. 1998. Personal Data Interchange.

International Standards Organisation. 2010. ISO 639-3 standard: SIL International.

Jorion, P. 2009. *Comment la vérité et la réalité furent inventées*. Paris: Gallimard.

Jullien, F. 1995. *Le détour et l'accès: stratégies du sens en Chine, en Grèce*. Paris: Grasset.

Jullien, F. 2009. *L'invention de l'idéal et le destin de l'Europe*. Paris: Seuil.

Kieser, A. 1997. Rhetoric and Myth in Management Fashion. *Organization*, 4(1): 49.

Korzybski, A. 1933. A Non-Aristotelian System and its Necessity for Rigour in Mathematics and Physics, *Science and Sanity*: 747–761.

Kuhn, T. S. 1961. The Function of Measurement in Modern Physical Science. *Isis*, 52(2): 161-193.

Kuhn, T. S. 1962. *The structure of scientific revolutions*. [Chicago]: University of Chicago Press.

Kuhn, T. S. 1996. *The Structure of Scientific Revolutions* (3 ed.): University of Chicago Press.

Labov, W. 2009. Volume 3: Cognitive and Cultural Factors, *Principles of Linguistic Change*.

Langer, S. K. K. 1948. *Philosophy in a new key, a study in the symbolism of reason, rite and art* ([1st ed.). New York,: Penguin Books.

Langer, S. K. K. 1953. *Feeling and form; a theory of art*. New York,: Scribner.

Lave, J., & Wenger, E. 1991. *Situated learning: legitimate peripheral participation*. Cambridge [England] ; New York: Cambridge University Press.

Lazer, D., & Friedman, A. 2007. The Network Structure of Exploration and Exploitation. *Administrative Science Quarterly*, 52(4): 667-693.

Lazer, D., Mergel, I., & Friedman, A. Co-citation of prominent social network articles in social science journals: Showing the change in research paradigm - how sociologists include references to science articles in their work.

Levinthal, D. A. 1997. Adaptation on rugged landscapes. *Management Science*, 43(7): 934-950.

Lewis, M. P. (Ed.). 2009. *Ethnologue: Languages of the World* (16th ed.). Dallas, TX: SIL International.

Lieberman, M. 2009. The directed graph of stereotypical incomprehensibility.

Lukacs, G. 1920. *History & Class Consciousness* (R. Livingstone, Trans.): Merlin Press.

Lyons, J. 1995. *Linguistic Semantics: an introduction*. Cambridge: Cambridge University Press.

Mahoney, T. A., Jerdee, T. H., & Carroll, S. J. 1965. The jobs of management. *Industrial Relations*, 4(2): 97–110.

Mandelbrot, B. B., & Hudson, R. L. 2004. *The (mis)behavior of markets: a fractal view of risk, ruin, and reward*. New York: Published by Basic Books.

Mangin, N. 2008. A comprehensive analytical framework for studying performance evaluation, *2008 Academy of Management Annual Meeting*. Anaheim, CA: Academy of Management.

Marcel, G. 2005. *Tu ne mourras pas*.

March, J. G. 1991. Exploration and Exploitation in Organizational Learning. *Organization Science*, 2(1): 71-87.

Marx, K. 1894. *Das Kapital*. New York: International Publishers.

Maslow, A. H. 1966. *The Psychology of Science: A Reconnaissance*. New York: Harper & Row.

Mead, M., & Bateson, G. 1976. For God's Sake, Margaret, *CoEvolution Quarterly*. Menlo Park, CA: Whole Earth Access Catalog.

Moreno, J. L. M. D. 1953. *Who Shall Survive?* (2nd ed.).

Moullet, M. 1992. *Le management clandestin*: Dunod.

Nash, J. F. 1951. Non-Cooperative Games. *The Annals of Mathematics*, 54(2): 286-295.

Nonaka, I. 1994. A dynamic theory of organizational knowledge creation. *Organization Science*, 5(1): 14.

Noronha, M. H. 2002. *Euclidean and Non-Euclidean Geometries*: Prentice Hall.

Obstfeld, D. 2005. Social Networks, the Tertius Iungens Orientation, and Involvement in Innovation. *Administrative Science Quarterly*, 50(1): 100.

OGC. 2008. KML.

OMA. 2000. WML.

Pamas, D. L. 1984. is ADA Too Big? *Communications of the ACM*, 27(11): 1155.

Peltokorpi, V. 2008. Organizational Transactive Memory Systems: Review and Extension.

Piaget, J. 1967. *Logique et connaissance scientifique*. Paris: Encyclopédie de la Pléiade.

Piaget, J. 1974. *Recherches sur la Contradiction*. Paris: Presses Universitaires de France.

Piaget, J., & Szeminska, A. 1941. *La genèse du nombre chez l'enfant*. Neuchâtel: Delachaux & Niestlé.

Plato. 1999. *Great Dialogues of Plato: Complete Texts of the Republic, Apology, Crito Phaido, Ion, and Meno, Vol. 1* (Warmington and Rouse ed.). New York: Signet Classics.

Platon. 1831. *Nomoi* (V. Cousin, Trans.): Pichon.

Platon. 1993. *Protagoras - Euthydeme - Gorgias - Ménexène - Ménon - Cratyle*: Flammarion.

Playfair, J., & Euclid. 1860. *Elements of geometry*: J.B. Lippincott & Co.

Polanyi, M. 1966. *The Tacit Dimension*. London: Routledge & Kegan Paul.

Popper, K. 2002. *The Logic of Scientific Discovery* (15 ed.): Routledge.

Porter, T. M. 1995. *Trust in Numbers*. Princeton: Princeton University Press.

Puffert, D. J. 2002. Path Dependence in Spatial Networks: The Standardization of Railway Track Gauge. *Explorations in Economic History*, 39(3): 282-314.

R Foundation for Statistical Computing. 2009. R: A language and environment for statistical computing. Vienna, Austria.

Ramírez, R. 1991. *The beauty of social organizations*. Munich: Accedo Verlagsgesellschaft.

Ramis, J.-P. 2009. Leonhard Euler, ou l'art de donner un sens à ce qui n'en avait pas. In S. M. d. France (Ed.), *Un texte, un mathématicien.*

Robinson, R. M. 1950. *An Essentially Undecidable Axiom System.* Paper presented at the International Congress of Mathematics, Cambridge 1950.

Roethlisberger, F. J., Dickson, W. J., & Wright, H. A. 1941. *Management and the Worker: An Account of a Research Program Conducted by the Western Electric Company, Hawthorne Works*: Harvard University Press.

Ronan, C., & Needham, J. 1978. *The Shorter Science and Civilization in China: An Abridgement of Needham's Original Text.* Cambridge: Cambridge Univerity Press.

Rosenberg, A. 1979. The Hardest Natural Languages. *Lingvisticae Investigationes*, III(2): 323-339.

Ryle, G. 1949. *The Concept of Mind.* London: Huchinson.

Schopman, J., & Shawky, A. 1996. Remarks on the impact of connectionism on our thinking about concepts. In A. Clark, & P. J. R. Millican (Eds.), *Connectionism, Concepts and Folk Psychology - The Legacy of Alan Turing*, Vol. 2. Oxford: Oxford University Press.

Schumpeter, J. A. 1962. *Capitalism, Socialism, and Democracy* (3rd ed.). New York: Harper & Row.

Scott, J. 2000. *Social network analysis - A handbook* (2nd ed.): Sage Publications.

Semay, C., & Silvestre-Brac, B. 2005. *Relativité restreinte : Bases et applications.* Paris: Dunod.

Simmel , G. 1902. Georg Simmel. "The Number of Members as Determining the Sociological Form of the Group: I. *The American Journal of Sociology*, 8: 1-46.

Simmel, G. 1910-11. How is society possible? *The American Journal of Sociology*(16): 372-391.

Simon, H. A. 1960. *The new science of management decision.* New York: Harper.

Spector, B. A. 2006. *Implementing Organizational Change: Theory Into Practice*: Prentice-Hall.

Stukeley, W. 1752. Memoirs of Sir Isaac Newton's Life. London: Royal Society.

Thorndike, E. L. 1918. The nature, purposes, and general methods of measurements of educational products. In G. M. Whipple (Ed.), *The Seventeenth yearbook of the National Society for Study of Education. Part II. The Measurement of Educational Products*: 16. Bloomington, IL: Public School Publishing Co.

Thorngate, W. 1976. "In General" vs. "It Depends": Some Comments of the Gergen-Schlenker Debate. *Personality and Social Psychology Bulletin*, 2: 404-410.

Tirole, J. 1988. Noncooperative Game Theory: A User's Manual *The Theory of Industrial Organization*: 479: The MIT Press.

Trepo, G. 1987. Introduction and diffusion of management tools: the example of quality circles and total quality control. *European Management Journal*, 5(4): 287.

Tsoukas, H. 1996. The firm as a distributed knowledge system: A constructionist approach. *Strategic Management Journal*, 17: 11.

Unesco. 2010. Translatium database: Unesco.

van Dam, A. 2008. Reception address. In Computer Graphics Pioneers (Ed.), *SIGGRAPH 2008.* Los Angeles, CA: Association for Computing Machinery.

von Foerster, H. 2003. *Understanding Understanding: Essays on Cybernetics and Cognition.* New York: Springer-Verlag.

von Hayek, F. A. 1945. The Use of Knowledge in Society. *American Economic Review*, 35(4): 519-530.

W3C. 1989. Semantic Web, Vol. 2010: World Wide Web Consortium.

W3C. 1999. HTML.

W3C. 2008. XML.

Wasserman, S., & Faust, K. 1994. *Social Network Analysis: Methods and Application*: Cambridge University Press.

Watts, D. J. 1999. *Small Worlds*: Princeton Univ. Press.

Watts, D. J., & Strogatz, S. H. 1998. Collective dynamics of small-world networks, *Nature*, Vol. 39: 440-442.

Weber, M. 1991. *Essays in Sociology* (1st ed.): Routledge.

Weick, K. E. 1995. *Sensemaking in Organizations*: Sage Publications.

Weick, K. E. 1999. Theory construction as disciplined reflexivity: tradeoffs in the 90s. *Academy of Management Review*, 24(4): 797-806.

Weick, K. E., & Bougon, M. 1986. Organizations as cognitive maps: charting ways to success and failures. In H. P. Sims, Jr., & D. A. Gioia (Eds.), *The thinking organization*: 102-135. San Francisco: Jossey-Bass.

Whitehead, A. N. 1979. *Process and Reality*: Free Press.

Williamson, O. E. 1995. Hierarchies, Markets and Power in the Economy: An Economic Perspective. *Industrial and Corporate Change*, 4(1): 21.

World Trade Organization. 2010. WTO statistics database.

Wurtz, O. 2009. The impact of personality on expatriate coping and cross-cultural adjustment, *2009 Academy of Management Annual Meeting*. Chicago, IL: Academy of Management.

XBRL Consortium. 2010. XBRL Specifications.

Xiao, J., & Li, J. 2008. *An Outline History of Chinese Philosophy* (S. Zhang, Trans.). Beijing: Foreign Languages Press.

43652 - mai 2012
Achevé d'imprimer par

1livre.com
du rêve à la réalité !

Made in the USA
Monee, IL
05 November 2026

67808942R10105